Problem Solving *for* Engineers

David G. Carmichael

CRC Press
Taylor & Francis Group
Boca Raton London New York

CRC Press is an imprint of the
Taylor & Francis Group, an **Informa** business

A SPON PRESS BOOK

CRC Press
Taylor & Francis Group
6000 Broken Sound Parkway NW, Suite 300
Boca Raton, FL 33487-2742

Printed on acid-free paper
Version Date: 20130220

International Standard Book Number-13: 978-1-4665-7061-0 (Paperback)

Library of Congress Cataloging-in-Publication Data

Carmichael, D. G., 1948-
    Problem solving for engineers / David G. Carmichael.
    pages cm
    Summary: "This book takes a systematic approach to problem definition, generation of
alternative solutions, analysis, and selection of the preferred solution. The book introduces
some fundamental terms needed to think systematically and undertake systematic problem
solving. It covers both individual and group problem solving. Selection of the preferred
solution involves decision making, and fundamental concepts of decision making are
introduced, including decision making in the presence of multiple criteria and uncertainty.
The treatment embodies decision making for sustainability, with its blend of economics, social
and environmental considerations. It also identifies and embodies the specific problem solving
of management and planning"-- Provided by publisher.
    Includes bibliographical references and index.
    ISBN 978-1-4665-7061-0 (pbk.)
    1. Project management. 2. Engineering--Decision making. 3. Problem solving. I. Title.

TA190.C373 2013
658.4'04--dc23   2013005343

Visit the Taylor & Francis Web site at
http://www.taylorandfrancis.com

and the CRC Press Web site at
http://www.crcpress.com

To my mother and father:

Edna K. Carmichael and A. Fraser Carmichael

Contents

Preface

Problem solving (implying decision making) is carried out every day by everyone. However, few people stop and think of the processes involved and whether they could improve their problem-solving or decision-making skills.

Doing the seminar circuit, there are a number of short public courses on the topic. Generally, these assume no prior knowledge and make little attempt to systematically present problem solving. Because they also assume no background of quantitative skills in systems-style thinking, they are unable to develop problem solving in any depth.

This book argues that the most rational way to develop a framework for problem solving is via a systems studies viewpoint.

Accordingly, the book first outlines *systems methodology*, *modeling*, and the various *systems configurations of analysis*, *synthesis*, and *investigation*. A systematic process is then outlined for problem solving. Problem solving and decision making are shown to lie within a systems synthesis configuration. Various forms of decision making are explored.

What Is a Problem and What Is a Solution?

In broad terms, a *problem* might be described as

being in a state different to that desired.

An alternative state to that existing is sought or wished for. A difference exists between what could or should happen and what is actually happening.

In broad terms, a *solution* might be described as

that which transfers the existing state to some other state.

There are degrees of goodness of solutions. Later, mention is made as to *how solutions might be ranked*. Mention is also made to *constraints* that restrict the choice of solutions.

In broad terms, a *state* is

> an indicator of behavior or performance.

Examples include the balance of a bank account, the health of a person, and the position and velocity of an aircraft.

These definitions of a problem and a solution are satisfactory for introductory or lay purposes but need tightening up for engineering purposes and are refined later in the book. More correctly, the state should be thought of and expressed as a variable that can take different values. Then a *problem* exists when

> the current values of the state variables are different to that desired

and the *solution*

> changes the values of the state variables.

The state variables remain the same from problem definition till after a solution is implemented. The only things that change are the values taken by these state variables.

Some examples might help clarify the intent of the meanings of problem, solution, and state.

Example 1. Problem: A bank account balance (state) is low or a higher balance (state) is desired. Possible solutions: Invest the money at a higher interest rate, deposit more money, etc.

Example 2. Problem: A person is unwell (state) or better health (state) is desired. Possible solutions: Take medicine, undertake exercise and a special diet, move to a sunnier climate, etc.

Example 3. Problem: A person is at location A (state) or desires to be at location B (state). Possible solutions: Drive by vehicle, catch public transport, walk, etc.

Example 4. Problem: A person is hungry (state) or desires not to be hungry (state). Solution: Anything that removes the hunger or transforms the person from being hungry to not being hungry.

These views on a problem and a solution are quite different to the majority of the literature and people's beliefs. Hopefully, by the end of this book, you will be persuaded to this style of thinking. The meaning of the term "state" is central to this understanding. Pay particular attention to this.

The approach presented in this book has developed out of existing systems engineering, systems studies, and systems theory thinking.

Dictionary definitions and lay usages of the term "problem" are rejected here as being unsuitable for developing a systematic framework for problem solving. You will also need to reject such definitions and usages. Typically, dictionaries talk of problems as "being something difficult, doubtful or hard to understand, there being degrees of severity of problems, and a problem being a matter requiring a solution."

Consistent with this, you will also need to banish from your thoughts the use of the term "problem" as encountered in textbooks and classrooms, meaning a contrived (for learning or entertainment purposes) "exercise," "question," or "puzzle," where a "solution" is sought by the text author or class teacher.

Exercise P.1

Some example situations that may occur in the workplace are

- A person continually turns up for work late.
- The wrong material is ordered.
- Equipment that has been ordered will not arrive on time; a delay is expected.
- Two people are incompatible.

All represent being in some undesirable state (that is, the values taken by the respective state variables are undesirable). The problem solution takes the situation to a more desirable state (that is, the values of the state variables are changed to something more desirable). A problem solution might be ranked in terms of goodness by how desirable the new values of the state are or the difficulty, complication, cost, etc., involved in arriving at a solution.

For each example, outline what you think are the state variables, the problem (in state terms), and possible solutions (that which converts an undesirable state to a more desirable state).

A problem may be categorized based on its cause (Figure P.1), but this may not be the most helpful way of looking at problems.

FIGURE P.1
Cause-based categorization of a problem.

Aim: The intent of the book is to present a rational and systematic approach to problem solving and decision making. The approach is intended to be nondiscipline specific.

The book presents a totally new approach to problem solving and relates this to existing treatments. The aims of the book are to understand problem solving and to contribute to thinking on problem solving. One of the overriding reasons for writing this book is to counter the myriad misconceptions and thinking errors that exist among writers on problem solving. The level of thinking that goes into problem solving in much of the literature is very superficial and cookbook in nature. To counter this, the book adopts a systems view to provide a rigorous framework. Rigor in the usage of terminology is also stressed. For technical terms, dictionary definitions and lay usage are rejected.

SOME LAY SAYINGS ON PROBLEMS

There is always an easy solution to every human problem: neat, plausible, and wrong.

H. L. Mencken

Don't fight the problem, decide it.

George C. Marshall

When ironing out problems, there is always the risk of putting in more creases.

P. K. Shaw

Comfortable problems are more acceptable than uncomfortable solutions.

Anon.

If at first you don't succeed ... you must have underestimated the extent of the problem.

Anon.

You cannot solve a problem with the same thinking that created it.

Popularly attributed to Einstein

htbhdydt

(An Australian colloquialism, meaning "how the bloody hell do you do that?")

Exercises

Exercises are set throughout the book to

- Make you think about and reinforce the material you have just read
- Test your understanding of the book material
- Make you think beyond the issues covered in the book
- Get you actively involved in the material

According to the Confucian saying:

> I hear and I forget
> I see and I remember
> I do and I understand

In many situations, there is frequently no right or wrong answer. In many cases, people are satisfied with a satisfactory outcome. The idea of an optimum solution may not exist. People from technical backgrounds may initially have difficulty in accepting that there is no right/wrong, black/white, on/off, yes/no answer.

Influential Publications

The following books by the author have ideas that were influential in the formation of the problem-solving ideas advanced in this book:

Structural Modelling and Optimization, Ellis Horwood Ltd. (John Wiley & Sons Ltd.), Chichester, U.K., 306 pp., 1981, ISBN 0 85312 283 0.

Engineering Queues in Construction and Mining, Ellis Horwood Ltd. (John Wiley & Sons Ltd.), Chichester, U.K., 378 pp., 1987, ISBN 0 7458 0212 5.

Construction Engineering Networks, Ellis Horwood Ltd. (John Wiley & Sons Ltd.), Chichester, U.K., 198 pp., 1989, ISBN 0 7458 0706 2.

Contracts and International Project Management, A.A. Balkema, Rotterdam, the Netherlands, 208 pp., 2000, ISBN 90 5809 324 7/333 6.

Disputes and International Projects, A.A. Balkema, Rotterdam, the Netherlands (Swets & Zeitlinger B. V., Lisse), 435 pp., 2002, ISBN 90 5809 326 3.

Project Management Framework, A.A. Balkema, Rotterdam, the Netherlands (Swets & Zeitlinger B. V., Lisse), 284 pp., 2004, ISBN 90 5809 325 5.

Project Planning, and Control, Taylor & Francis, London, U.K., 328 pp., 2006, ISBN 0 415 34726 2.

Author

David G. Carmichael is professor of civil engineering and former head of the Department of Engineering Construction and Management at the University of New South Wales, Kensington, New South Wales, Australia. He is a graduate of the Universities of Sydney and Canterbury; a fellow of the Institution of Engineers, Australia; a member of the American Society of Civil Engineers; and a former graded arbitrator and mediator.

Dr. Carmichael publishes, teaches, and consults widely in most aspects of project management, construction management, systems engineering, and problem solving. He is known for his left-field, nonmainstream thinking on project and risk management (*Project Management Framework*, A.A. Balkema, Rotterdam, the Netherlands, 2004), project planning (*Project Planning, and Control*, Taylor & Francis, London, U.K., 2006), and most other matters involving problem solving.

1

Systems Methodology

1.1 Introduction

Systems thinking (equivalently, systems approaches, engineering and studies) gives a methodology for dealing with problems, not only technical and organizational problems but also social, political, environmental (natural environment), and others problems. The methodology is essentially based on generalizations of real cases. Systems thinking involves abstraction—the process of formulating generalized ideas or concepts by extracting common qualities from specific examples. It integrates disciplines and offers unifying principles and a common language. Its goal is to provide an all-encompassing treatment of systems (but that goal has only partly been reached at the present day). It breaks down the jargon barrier between disciplines (though in doing so, it necessarily introduces new jargon). It is a holistic approach that looks at the interaction of the system with all outside the system (referred to as the environment, a term not to be confused with the natural environment), and the interaction of the system components. With large-scale, complicated, and interactive systems, it becomes multidisciplinary.

The term "systematic" is often merged with "systems thinking." This is true in the sense that "systematic" means ordered and structured.

Systems, as a discipline in its own right, has attracted some highly impressive thinkers such that a significant body of concepts supported (and, in some cases, oversupported) by a body of mathematics now exists. Unfortunately, too many people emphasize the mathematical side at the expense of the conceptual side. It is the author's belief that the conceptual side gives the power to problem solving. The mathematics helps, but it should come second, not first. The mathematics should only be used if it helps, rather than it being the focus.

This discussion, of course, assumes the reader has some notion of what a system is. It is a much-abused word. A system is defined in detail later.

Outline: The following sections attempt to explain the terminology and concepts related to systems thinking.

1.2 Terminology

The term "system" or "systems" arises in many places. Apart from its lay or dictionary usage and its overusage (and abuse) by technical and non-technical people in a lay or imprecise sense, it arises in connection with discussion in

- General systems theory
- Systems theory
- Systems thinking
- Systems science
- Systems methodology
- Systems approach
- Systems analysis, synthesis, investigation
- Systems engineering
- Systems studies

and related areas.

There is an overlap among all of these areas dealing with systems. Ideas are borrowed freely and mixed. Together, a reasonably coherent set of concepts and thoughts can be developed and supported by quite sophisticated (and sometimes overly extravagant) mathematical models.

A systems approach by its nature is not discipline dependent, and it is not restricted to a person's degree of vocational training. It gives a fundamental framework to problem solving and decision making.

In some disciplines, the term "process" may be used synonymously with the term "system," but the term process is also used in other senses. More usually, process refers to a series of transformations such as in a production process. To add confusion, a production process, for example, may be referred to as a system. In this book, the term "process" is not used synonymously with "system," but rather more in the sense of being a transformation or sequence of steps.

1.2.1 Abuse of Terminology

Unfortunately, as mentioned, the term "system" is overused and overabused. When people cannot think of a word, they lazily call it a system. This book, by contrast, adopts a consistent usage of the term from start to finish. Similarly, later in the book, when terms such as "model," "control," "state," and "output" are used, these are used consistently in the defined sense and not in any lay person's or dictionary sense. Unfortunately, lay and nonlay persons and many publications use these terms any way they want to. As with *Alice in Wonderland*,

'When I use a word,' Humpty Dumpty said in a rather scornful tone, 'it means just what I choose it to mean, neither more nor less'. 'The question is', said Alice, 'whether you can make words mean so many different things'.

Through the Looking Glass, **Lewis Carroll**

In this book, a consistent and tight usage for all terminology is followed.

It is always intriguing to know how people can communicate on systems, modeling, and the like when they do not have well-defined meanings for these terms. For a poet or a novelist, this is a bonus; but not for technical discussion and advancement of the state of the art in any discipline. It is like people conversing with each other (without the body language) in foreign languages; clearly, nothing is being absorbed or transferred from one person to another. The management literature is full of such terminology imprecision (Carmichael, 2004, 2006), and as a consequence, the state of the art of management is not progressing.

1.2.2 Probabilistic, Deterministic

Two terms that arise frequently are probabilistic (or stochastic) and deterministic. Probabilistic refers to something, for example, the system behavior, being described in terms of frequencies of occurrence or in terms of probabilities; it implies some variability or uncertainty. Deterministic, on the other hand, implies certainty; for example, the system behavior is known with certainty and is predictable to something definite.

Deterministic variables are commonly described in terms of their mean, average, or expected values. Probabilistic variables are commonly described in terms of probability distributions or, if using a second-order moment approach, in terms of expected values and variances (standard deviation squared); the variances capture the variability information or uncertainty.

Risk, for example, only exists in the presence of uncertainty, and hence risk approaches are probabilistic. With certainty, there is no risk.

Generally, determinism is simpler to deal with, and, wherever possible, people simplify from probabilistic to deterministic approaches.

1.2.3 Dynamic, Static

The terms dynamic or nonstationary refer to something that changes with time. The terms static or stationary refer to something that does not change with time.

1.2.4 Discrete, Continuous

The term "discrete" implies that time (or spatial coordinates) is only defined at certain points. Hence, a discrete system is one in which the system variables change only at these discrete points in time or between points in time. The term "continuous" implies that time (or spatial coordinates) is defined everywhere. Hence, a continuous system is one in which the variables change continuously over time.

For example, consider water flowing into a dam from rainfall and water being drawn off for irrigation. The head of water in the dam is changing continuously over time. If, however, observations of the water level were only made, say, daily, the head of water is changing in a discrete fashion.

The distinction between continuous and discrete becomes apparent if you look at the distinction between differential equations (continuous) and difference equations (discrete).

Exercise 1.1

Do you agree that it is possible to have a body of knowledge that encompasses all disciplines? Would such a body of knowledge lack content at the expense of generality, or is it possible to balance content and generality?

1.3 Origin

When did the idea of systems thinking begin? Its origins are said to be in a number of sources, but generally in a quest to understand both natural and man-made systems, and a belief in a commonality of these. Broadly, systems thinking stems from the following:

- An acknowledgment that complicated situations cannot be dealt with unless a person happens to be very clever. People tend to prefer simple situations rather than the real, and possibly complicated, situations.
- People wish to examine the behavior of complicated systems made up of interacting components. Systems referred to here not only include electrical–mechanical systems but also social systems, biological systems, economic systems, etc., and combinations of these.

Early well-publicized influences on systems thinking include

- von Bertalanffy (1920s, 1930s)—general systems theory and "organismic" biology
- Whitehead (1920s)—organic mechanism

- Cannon (1920s, 1930s)—homeostasis
- "Gestalt" psychology (1920s)—stimulus–response (robot) models of human behavior
- Holistic approaches such as in "The Limits to Growth, a Report for the Club of Rome, Project on the Predicament of Mankind" (1970s), where the interrelationship of parts (here population, food, etc.) was emphasized
- Kohler (1920s)—Gestalten (wholes) in physics
- Lotka (1920s)—open systems, analogies between organic and inorganic systems, system–environment interaction
- Szilard (1920s)—entropy and information relationship
- Wiener (1940s)—cybernetics, feedback
- Shannon and Weaver (1940s)—information theory
- Beer (1950s)—cybernetics and management
- Ackoff and Churchman (1950s, 1960s)—operations research
- Simon (1950s)—organizations

However, note that there have been thousands of other contributors to systems thinking, and these contributors are rarely rewarded in the literature in terms of acknowledgment of their work.

The fields of operations research and systems studies considerably overlap, and it can be more a case of which discipline a person comes from that determines the field allocated for the respective contribution.

> **Exercise 1.2**
>
> How might you think a discussion on systems might be connected with problem solving? Or a background in systems studies might improve your understanding of the problem-solving process?

1.4 System

A *system* is composed of *interacting* or interrelated entities, elements, objects, parts, or components (equivalently, *subsystems*) (Figure 1.1).

Whatever does not belong to a system is referred to as the *environment* (and should not be confused with the natural environment). This automatically establishes the system boundary. Discussion on the environment and choice of the system boundary is given in the following text.

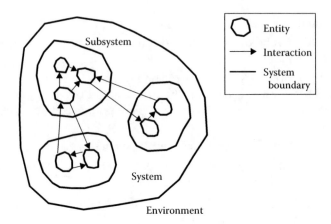

FIGURE 1.1
Representation of a system.

External to the system, *inputs* and *outputs* take place. The nature of the inputs and outputs and their characterization depend on the system study being undertaken. In this sense, a system may also be thought of as an *input–output transformation*.

The distinction is sometimes made between a system and an aggregate. An aggregate is composed of parts like a system is composed of parts, but the parts bear no relationship. The analogy with letters of the alphabet might be drawn. Letters together, if they form a word, might be regarded as analogous to a system, but analogous to an aggregate if they do not form a word; the connection of the parts is important here.

Because the parts interrelate, so the properties of the system are different to that of the parts. Entities or parts at their lowest level need not be defined—at that level, they can be regarded as primitives. However, they do have properties or attributes, and these properties can be measured and take on values. In studying any system, some properties or attributes might be selected as relevant to the study, and other properties may be ignored. It will depend on the intent of the study.

1.4.1 Relation between Parts

The relation between parts describes the way parts are dependent on each other or, more correctly, the way the properties of parts are dependent on each other. Mathematically, a relation implies a function. If a change in a property of one part changes a property in another part, then a relation exists.

The relations between the parts influence the system behavior.

In studying any system, some relations might be selected as relevant to the study and other relations may be ignored. It will depend on the intent of the study.

Exercise 1.3

Consider a company's organization. Identify the system and parts (subsystems), and relationships between the subsystems.

Exercise 1.4

From the viewpoint of a structural engineer, a structure is composed of members, which are composed of elements, which are composed of materials. Equilibrium and compatibility relationships represent subsystem interaction while constitutive relationships represent subsystem models of behavior. (See Figure 1.2, and Carmichael, 1981.)

Why do structural engineers generally not go below the level of material characteristics? Why are they generally content only with macromaterial behavior?

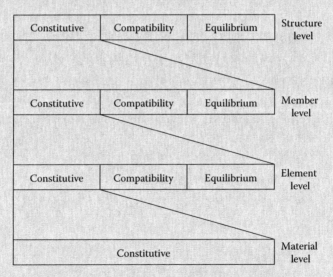

Constitutive	Compatibility	Equilibrium	Structure level
Constitutive	Compatibility	Equilibrium	Member level
Constitutive	Compatibility	Equilibrium	Element level
Constitutive			Material level

FIGURE 1.2
Relationship between levels. A system on a given level is a subsystem on the next higher level.

1.5 Fundamental Variables

Certain fundamental system variables can be recognized. These include *input, output, state,* and *control.* There is also what is called *disturbance.* They all have synonyms in the systems literature, used variously by different writers. All terms are used very loosely by people generally. Tight meanings are given in the following text. Banish from your mind all meanings and usages other than those given as preferred here.

1.5.1 Output, State

A system *output* describes the external behavior (performance, response) of a system. This is the behavior that is observed and measured. If the system is a dynamic system, this behavior or performance changes over time.

An internal descriptor of behavior is *state.* The system's state is nonobservable and nonmeasurable. For a dynamic system, it will be found convenient later to regard the present state as containing a record of the system's past states, as well as the present condition.

In many cases, the state and the output turn out to be, or are assumed to be, the same. There is a one-to-one relationship between output and state. (Sometimes this is expressed in control systems theory, through what is called an observation equation, where "output = state" [Carmichael, 1981, discussed later]. And there are no "observability" issues in the sense of Kalman [also discussed later]. Observations and measurements are only possible on the system output, and from this, the state has to be inferred.)

Example

Consider a bank account. The current balance is the state, which reflects deposits and withdrawals (inputs). Further deposits and withdrawals, together with knowledge of the current balance, will give the new balance. Here the account balance is both the state and the output.

Example

In project planning, through the way projects are conventionally represented in a "stripped-back" form, the internal and external behavior of projects are the same.

The most obvious states for a project are the resources (or money) used up to any point in time (cumulative resources/cumulative cost) and the production up to the same point (cumulative production).

The behavior of system parts may also be described in terms of states, expressed at the level of the parts.

[Conventionally, relations between parts do not have states, but could be interpreted to have relations if regarded as parts themselves. It is even possible to reverse the role of a system part and a relation between parts, in order to study the system in a different way. However, if this confuses more than it helps, then ignore this comment for now and return to it when you have mastered the later chapters.]

Example

As an example of reversing a system part and the interaction between system parts, consider the reverse forms of activity-on-arrow and activity-on-node diagrams that are used in the critical path method (project planning). They can be shown through fairly involved mathematics that they are equivalent. In general terms, the roles of nodes and links of a network are interchanged. (See Figure 1.3, and Carmichael, 2006.)

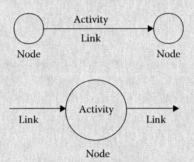

FIGURE 1.3
Activity-on-arrow and activity-on-node network diagrams, respectively.

A system could have many possible descriptors of state. Depending on the intent of the system study, and/or the relative importance of the states, certain system state descriptors might be examined closely and the rest ignored.

Exercise 1.5

People often talk of "state of health," "state of a country's economy," etc. Are these usages of the term "state" consistent with the earlier descriptions?

Exercise 1.6

To make you think about what the term "state" means, answer the following question: The Mississippi River flows in which state?

1.5.2 Input, Control

Systems also have inputs. An input might be classified in terms of whether it is chosen, selected, or influenced by the engineer (or caused to be selected or influenced by the engineer), or whether it is not. The former input is referred to as a *control*, decision, or action (or design variable), depending on the application and the writer. The latter input is not given any special name and is just referred to as an *input*; typically, such inputs are beyond the influence of the engineer.

Generally, usage of the term "control" is favored in this book in place of decision and action.

Example

In project planning, through the way projects are represented in a "stripped-back" form, the most obvious choice of controls are work method, resource (people and equipment) selections, and resource production rates. Work method, resources, and resource production rates may be freely selected by the planner, subject to any constraints on this selection being present, in order to get the project to where the planner wants it to go (in terms of cost and schedule).

When discussion later turns to the synthesis configuration and, in particular, practices of optimization and decision making, the preferred controls will be seen to be those that *extremize* a defined objective(s). Extremize means to either *minimize* or *maximize* (generally, *optimize*) as the case may be. The number and type of controls available to the engineer may be restricted by what are referred to as *constraints*.

Generally, the term "control" is used loosely in the literature and by lay persons and practitioners. The earlier quote of Lewis Carroll is particularly relevant here. The term is used almost as loosely as the term "system" is. Sometimes the term is used just because it sounds good or impressive, but with no real meaning in mind. In many cases, and particularly in a management or planning context, the term is used to mean placing some limit or containment on something such as costs, or keeping this something from growing unacceptably. In general management texts, and lay usage, the term "control" is used in the context of power, dominance, or influence. (Something that is "out of control" is unable to be influenced.)

In this book, the term "control" is used in a systems sense, meaning *the actions, input, or decisions taken in order to bring about a desired system behavior (performance)*. Banish from your mind all other dictionary and lay usages.

In the synthesis configuration discussed later, the popular lay usage of control as stopping something from growing unacceptably, or containing something, is incorporated through the use of "constraints."

1.5.3 Disturbance

In some systems studies, there is something present that prevents the hoped-for output being obtained or measured exactly, or prevents the state from being established exactly. Commonly, this is because of an uncertain changing environment. Collectively, this corruption is referred to as *disturbance*, though control systems texts may use the term "noise".

1.6 Subsystems

A subsystem is a system and is part of a larger system or a larger whole. Entities, parts, and components may be regarded as subsystems.

In studying systems, systems are decomposed to lower-level subsystems, stopping at an appropriate resolution level (Figure 1.4). The intent of the study will determine to what resolution the decomposition is taken.

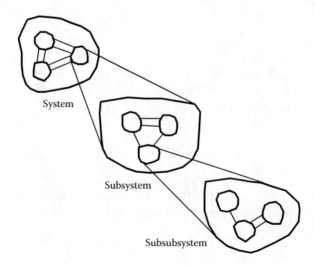

FIGURE 1.4
Different resolution levels of a system.

Example: An Organization

Organizations have attracted much study, even though the state of the art of knowledge about organizations is quite crude. An organization is made up of interdependent parts interacting with each other and the environment.

An organization might be regarded as a hierarchical multilevel system with a tree of ordered subsystems (Figure 1.5).

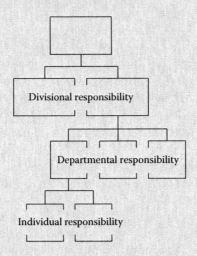

FIGURE 1.5
Example organizational hierarchy.

Example: Framed Structure

A structural engineer regards a framed structure as being composed of beams, columns, and slabs. The beams, columns, and slabs are subsystems of the framed structure (system).

Example: Project

A project is comprised of subprojects, which are comprised of sub-subprojects, which are comprised of ..., right down to the activity or task level (Figure 1.6).

```
                    Project
                    1000  subproject
                      1100  activity/task group
                      1101  activity
                      1102  activity
                      1103  etc.
                      1200  activity/task group
                      1201  activity
                      1202  etc.
                      1300  etc.
                    2000  subproject

                    3000  etc.
```

FIGURE 1.6
Example hierarchical breakdown of a project.

Example: Book

A book is composed of chapters, which are composed of paragraphs, which are composed of punctuation and sentences, which are composed of words, which are composed of letters (Figure 1.7).

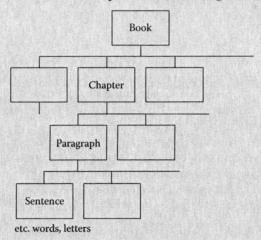

FIGURE 1.7
Hierarchical breakdown of a book.

Exercise 1.7

Programs representing planning information are often presented with different levels of detail. People at the workface require much detail, and the time horizon is short. People in the project team require less detail and a longer time horizon. Senior staff require only broad information and large time horizons.

Represent such programs in terms of a hierarchy of subsystems.

1.7 Environment

Whatever does not belong to a system is referred to as the environment. This automatically establishes the system boundary.

When interactions or interchanges occur between the system and the environment, that is, across the system boundary, the system is referred to as *open*. When such interactions or interchanges do not occur, the system is referred to as *closed*. The interaction or interchange may be either way or both ways.

Example: Climate and People

The climate influences people, but individuals, by and large, do not influence the climate (ignoring carbon emissions).

The choice of system boundary, and hence the choice as to what is considered the system and what is not considered the system (that is, the environment), will depend on the purpose of the study. Anything at all, in principle, can be regarded as the system, but it is not arbitrarily chosen—rather it is chosen with the intent of the study in mind.

The environment might be regarded as a supersystem of the system.

In this book, when reference is made to the nature or ecological version of environment, it is referred to as the "natural environment" in order to distinguish it from this system's use of the term "environment." Both usages mean "surroundings," but their intent is different.

Exercise 1.8

In a study of mechanical plant or equipment productivity, what would be reasonable to choose as the system and what would be reasonable to choose as the environment?

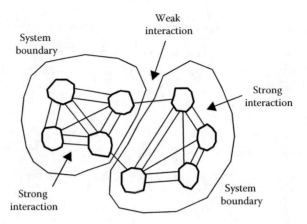

FIGURE 1.8
Selecting the system boundary.

1.8 System Boundary

The boundary is what divides the system and the environment. The boundary may be something physical but need not be.

Example: An Organization

When studying a company, the boundary might be the physical walls of the building housing the company. This might be useful for studying inventories, deliveries, and dispatches. However, such a boundary is not relevant if examining the financial aspects or the human relations aspects of the company.

The choice of the boundary depends on the intent of the study.

Consideration is often given to selecting a system boundary, where the level of interaction between the system and the environment is minimal or less concentrated. Higher concentrations are incorporated into or excluded from what is regarded as the system. That is, internal relations are strong. External relations are weaker. The notion of clustering is relevant here (Figure 1.8).

1.8.1 Open and Closed Systems

Generally, systems are open (that is, interactions occur across the system boundary), although for simplification they may be regarded as partly closed. Some or all interactions between the system and the environment might be ignored in order to make the study more tractable. In a sense, this technically changes the definition of what is the environment.

Example: Thermodynamics

The classical laws of thermodynamics are based on assumptions of closed systems. Energy exchanges are only considered within the designed system.

Example: Economics

Some economic studies assume no money goes beyond set boundaries.

Exercise 1.9

Consider a company's organization and its division into departments. How do you rank the strength of interaction within a department compared with interaction between departments? Accordingly, would appropriate boundaries be those that encompass each department?

1.9 System and Behavior Characterization

A number of characterizations of dynamic systems and their behavior (output, state—performance, response) may be distinguished. Of interest with many dynamic systems is the behavior with respect to

- Equilibrium
- Steady state
- Transient state
- Equifinality
- Stability

The term "process" may be used to describe a series of states that something or a system goes through, as in the transformation that occurs, for example, in a manufacturing or production process.

1.9.1 Transient and Steady States

A transient state is one that changes with time. A steady state stays constant with time. A dynamic system, after it starts or is perturbed, will behave transiently before perhaps entering a steady state.

Example

When injured or during an illness, a human's body temperature rises, only returning to a relatively unchanging temperature when well. This is achieved through metabolism and heat exchanges with the environment.

Example: Earthmoving or Traffic

At the start of a shift, a fleet of trucks involved in earthmoving may all be queuing at the loader. Only after all trucks have been loaded and gone on their haul routes does the system settle down to a steady state.

At traffic lights, vehicles queue waiting for the green light. It is not until some distance after the traffic lights that vehicles flow in a steady fashion.

Example: Management Fads

One of the by-products of management fads is that they perturb the steady state, making people work differently and perhaps more effectively than before. After a while, however, another steady state returns, which requires a further fad to perturb it. Are these fads, then, cunning managerial solutions designed to lift worker productivity through perturbation rather than through anything intrinsic in the fad itself? Perhaps something like Figure 1.9 applies.

There is an argument that organizations introduce a management fad as a way of enthusing the workforce. After a while, the transient effect disappears and the organization settles into a steady state, possibly of perceived lower productivity. In order to raise the productivity, another transient is introduced in the form of another management fad whose effect after a while also disappears. Thus, the logic is to never let the organization reach a study state, but rather to keep it in a continual transient state of perceived higher productivity.

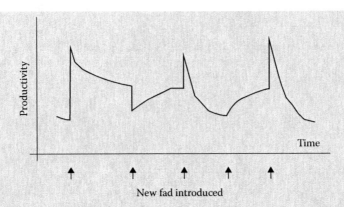

FIGURE 1.9
Speculated effect of fads.

Similar observations are made on the introduction of incentives; these have to be continuously changed else the workforce regards them as the norm. Nine-day fortnights, flexible hours, etc., are well received initially, but their effect disappears with time. Consider the traditional tea/coffee break and its influence on production throughout a working day.

There is a sort of Hawthorne effect operating here (people responding to the attention, rather than the managerial solution).

1.9.2 Equilibrium

Equilibrium is a term used to describe a steady state, but usually refers to a closed system, that is, when there is no interaction with the environment.

1.9.3 Equifinality

When a system reaches the one final state from different initial conditions, it is referred to as equifinality. This is only possible with open systems because of their interaction with the environment. The final state of closed systems is determined by their initial conditions.

1.9.4 Stability

Stability refers to a system adopting a certain state with time after a stimulation is removed. In some cases, the system may oscillate and converge or return to its initial state; in other cases, the system might tend toward some other state.

Stable systems have a tendency or preference for a desired state—they have a property of finality or teleology.

Example

If a pendulum is stimulated, it will eventually return to its initial position because of frictional forces (Figure 1.10).

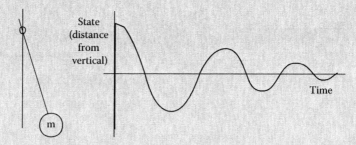

FIGURE 1.10
Behavior of a pendulum.

Example

In economics, price may fluctuate depending on the supply and demand relationship. The cobweb in Figure 1.11 shows an unstable situation.

Oscillations such as shown in Figure 1.10, should they move further away from the horizontal axis or not converge, would indicate instability.

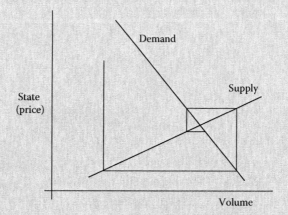

FIGURE 1.11
Cobweb.

A system may be stable to only certain stimulations and not others. Ultrastable systems are stable to all stimulations.

So-called *closed-loop systems, adaptive systems,* and *self-organizing systems* tend to be stable.

Exercise 1.10

Consider your work behavior over a day, over a week, and over a year. Identify the transient and steady-state periods.

Consider your work behavior in starting a new job, perhaps in a new organization. Identify the transient and steady-state periods.

2

Models and Modeling

2.1 Introduction

> We think in generalities, but we live in detail.
>
> **A. N. Whitehead**

Core to a systems approach to solving problems is the notion of models. Models are essential aids in problem solving.

A model is a representation or image of the system. It is used to examine alternative systems and behavior because it is easier to study than the system itself. The model contains information about the system.

Any system can have an infinite number of models. Some will be better than others for the purpose of the system study.

> The choice is always the same. You can make your model more complex and more faithful to reality, or you can make it simpler and easier to handle. Only the most naive scientist believes that the perfect model is the one that perfectly represents reality. Such a model would have the same drawbacks as a map as large and detailed as the city it represents, a map depicting every park, every street, every building, every tree, every pothole, every inhabitant, and every map. Were such a map possible, its specificity would defeat its purpose: to generalize and abstract.
>
> Map makers highlight such features as their clients choose. Whatever their purpose, maps and models must simplify as much as they mimic the world.
>
> **Gleick, 1988**

A model is not the real thing. It is artificial. The question is whether the model captures some piece of reality in a reasonable way. Its use depends on its appropriateness. One principle of modeling is *less is more*. That is, models are all about what reality can be left out (and still get interesting insights), not what can be fitted in (and still be able to handle the model). This is a principle that takes time to appreciate.

A model is an image of a system and assists in the study of the system. The purpose of a model is that it is easier to study than the system itself. Being an image, it must bear some resemblance to the system.

Where the study is in an area where little knowledge exists, a model (theory) from some other area may be borrowed and analogies drawn. The model is then used to develop knowledge in this new area. Where the system appears very complicated, a simplified mathematical relationship might be tried based on simplified assumptions. Where direct experimentation of a system is too expensive, dangerous, or difficult, experimentation with a model is more suitable. Where the system is not physical and direct observations are difficult, a model may help.

> **Example**
>
> In the evolution of science, models describing physical systems are put forward, perhaps criticized and forgotten, perhaps accepted, and maybe improved upon. Consider the explanation of magnetic and electrical phenomena, and the advancement of medicine.

Given that there is uncertainty associated with most systems, it would appear that most models should be probabilistic. However, this may make the models too unwieldy to use, and any associated manipulation too complicated. Instead, most modeling opts for more simplified deterministic models.

> **Exercise 2.1**
>
> Under what circumstances might you use deterministic models, and under what circumstances might you use probabilistic models?

> **Exercise 2.2**
>
> People are aware of most system behavior having variability, yet we continue to do our calculations deterministically. Is this logically consistent or justifiable? How big a simplification is it to ignore variability?
>
> In terms of uncertainty, there are known knowns—things we know we know. There are known unknowns—we know there are some things we do not know. There are also unknown unknowns—the ones we do not know we do not know. How does this contribute to how variability is handled?

2.1.1 Terminology

The terms *model*, *theory*, *hypothesis*, and *law* might be used interchangeably by some people. More usually, hypothesis is something to be tested, and demonstrated to hold or be refuted, while the others may or may not have been already demonstrated or refuted. Some people use the term "theory" (and even "model"), particularly in the popular management literature, implying that the proof is self-evident or not necessary. That is, the theory is never proven, but rather is held out to be a compelling truth, even though it may not be. Discussion on the "scientific method" later picks up on these issues.

There is also abuse of the term "model." People use the term in the sense of *Alice in Wonderland*, to mean anything that they want it to mean.

Exercise 2.3

How else could models be used to understand systems?

Outline: The intent of the following sections is to fully explore why models exist, their usage, and their benefits. Examples of models from various disciplines are given.

2.2 Formalism

Formally what happens, in using models, is that there are two systems:

1. That being studied S.
2. Its model M.

M is used to obtain information about S.

Where S and M are isomorphic (or same form), they have corresponding structures, and the pattern of relations is the same. There is a one-to-one mapping of entities and relations between S and M (Figure 2.1).

Generally M will be simpler than S, and some relations and entities present in S will be omitted in M. That is, there is a many-to-one mapping between S and M. S and M are homomorphic (Figure 2.1). Homomorphic mapping is only one way.

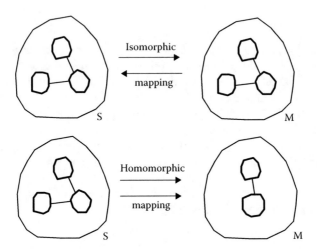

FIGURE 2.1
Isomorphic and homomorphic mappings.

Example

Electrical analogues of mechanical systems such that mass, frictional force, and stiffness might be replaced with inductance, resistance, and capacitance in an electrical system or an associated mathematical equation—the mechanical system, electrical system, and formal equations may be regarded as models of each other.

This is an example of isomorphism. Everything in the mechanical system is reproduced with something equivalent in the electrical network.

Exercise 2.4

There are two main kinds of clocks—analogue and digital. The usual analogue clock has hour, minute, and second hands. The digital clock only has numbers. Are these isomorphic or homomorphic? How would you classify a sundial?

Exercise 2.5

A photograph is a model used to represent some scene at a point in time. Is the relationship between the photograph and the scene isomorphic or homomorphic? If it is isomorphic, what changes would convert it to being homomorphic? If it is homomorphic, what changes would convert it to being isomorphic?

2.3 Hierarchical Multilevel Systems

The concept of a system provides for system representation as an assemblage of subsystems. As such, any system may be viewed as a hierarchical multi-level system with subsystems (Figure 2.2).

The use of the terms "higher" and "lower" when referring to levels or strata is interpreted in the sense of the orientation or hierarchical decomposition of Figure 2.2.

States, controls, and outputs exist at all levels.

No levels are isolated; when considering any one level, the two adjacent levels must be taken into account. Information from higher levels is needed to solve lower-level problems; upper levels define the bounds within which the lower levels function. The construction (and behavior) of the higher levels depends on the lower-level construction (and behavior). Control may be applied, and exchanges with the environment (everything outside the system) may occur at all levels. Changes in controls on higher levels are manifested by parameter changes on lower levels. Understanding of the system functioning improves on ascending the hierarchy, while the detail unfolds on descending the hierarchy. Explanations of the total system behavior are possible in terms of the lower levels and their interrelationships.

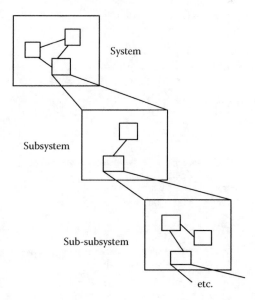

FIGURE 2.2
Multiple levels of subsystems.

Example: Structures

The subsystems for a building structure would generally be chosen to correspond to (in descending levels) the structure, member, element, and material levels. Alternatively, substructures, member segments, cross-section layers, and related ideas may be interpreted as subsystems.

Each subsystem model is in terms of a constitutive relationship, while subsystem interaction is in terms of equilibrium and compatibility relationships. The three sets of relationships, when combined, define the subsystem description at the next higher level. That is, on any given level, the behavior is studied in terms of that level's constitutive relationship, while the manner in which subsystems on that level interact to form a higher level system is studied on the higher level.

Example: Project

For understanding projects, it is convenient to build up the fundamental relationships from the lower levels, referred to here (from lowest level upward) as constituents, elements, and activities (Figure 2.3).

Constituents feed up to elements; elements combine to give an activity; activities combine to give a project (and projects can be combined to give a "program").

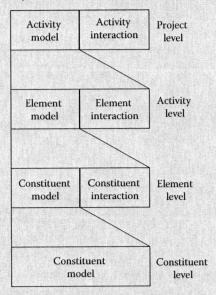

FIGURE 2.3
Building up a project model.

Example: Organizations

Consider an organization. The various levels might be as shown in Figure 2.4. The first division of the organization might be called a department, division, line, or function. Some modern management literature might suggest that the first division be according to a process rather than a department; it still remains multilevel nevertheless. Some modern management literature might also suggest that organizations can exist without demonstrating any multilevel qualities, but this is an unstable situation—the organization takes on a hierarchy despite the best of attempts by people to do away with hierarchies.

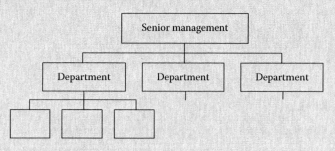

FIGURE 2.4
Organizational structure example.

Example: Taxonomy

In taxonomy, organisms are classified hierarchically:

    Kingdom
        Phylum/Division (Subphylum)
            Class
                Order
                    Family
                        Genus
                            Species

The hierarchy reflects the evolution of the species, deduced from morphological and physiological similarities between animals/plants.

An animal/plant is usually named in a binomial way (Genus Species) after Carolus Linnaeus.

Example: The common dog (*Canis familiaris*):

Kingdom: Animalia
Phylum: Chordata
Class: Mammalia
Order: Carnivora
Family: Canidae
Genus: Canis
Species: Familiaris

Example: Family Tree

The family tree, starting at two persons, expands as later generations are added. The tree progresses over generations with the subsystem level changing in name only as new generations are added: ..., great grandparents, grandparents, parents, child, ...

Example: Books

A book is divided into chapters, sections, subsections, paragraphs, sentences, words, and letters. Each is a subsystem of the next higher level. Each subsystem combines with like subsystems to produce the next higher level.

Example: Work Breakdown

A project (scope) can be decomposed into subprojects and then into sub-subprojects, and so on down to the activity level. Alternatively, intermediate subsystems might be viewed as subprojects and work packages. The lower levels represent finer detail. The process of decomposing a project (scope) might be called scope delineation (or scope definition) by some writers. The term "work breakdown structure" (WBS) is used to refer to the resulting systematic decomposition.

Work breakdown is not unique; various work breakdowns are possible for any project. A breakdown is chosen that makes life as easy as possible for the planner. There is no right or wrong answer to this, only better or worse subdivisions. Irrespective of the breakdown adopted, the collective bottom level of activities should be the same. All the work that has to be done to complete a project is contained within the WBS.

Terminology for each of the levels, used by different people, might include project, subproject, work package, job, subjob, process, or activity. The terminology of task and subtask might also be used at the middle to lower levels. There is no consensus in the use of terminology to describe the various levels, though the terminology "project"—"subproject"—...—"activity" is popular.

What is not evident from the WBS is the degree of interrelationship between its components, or the reliance of some on others, and how the WBS has been used to separate individual components.

Work breakdown may be represented as a tree diagram (Figure 2.5), or appropriately indented dot points, or appropriately numbered items (Figure 2.6) that recognize the hierarchy involved.

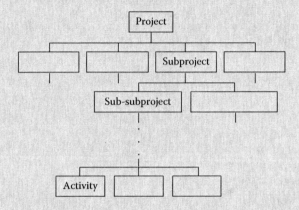

FIGURE 2.5
WBS (tree diagram form).

```
Project
1000  Subproject
 1100  Sub-subproject
  1101  Activity
  1102  Activity
  1103  etc.
 1200  Sub-subproject
  1201  Activity
  1202  etc.
 1300  etc.
2000  Subproject

3000  etc.
```

FIGURE 2.6
Example hierarchical breakdown and numbering of activities.

With respect to Figure 2.6, there is no set standard for numbering schemes, or numbering for accounting purposes. Rather, the schemes may reflect different project characteristics and different organizational ways of dealing, for example, with money.

Subprojects may be chosen at the convenience of the planner, but might follow a breakdown according to

- Phase (chronological): time period phasing
- Work type or nature: resource groupings
- Work parcels: the way the work is to be carried out
- Contracts
- Region, location, or geographical area
- Organizational breakdown
- (Existing) cost codes, cost centers, cost headings: project cost and data summary reports
- The nature of the planning network: activity groupings
- Function

Lower levels can be integrated or aggregated (folded up) to higher levels for reporting or communication purposes. Any level is the sum of the immediate lower-level work.

With this aggregation comes an aggregation of duration, resource, and cost information, such that information on durations, resources, and cost can be given at each level. Monitoring (project production/progress and resource usage), reporting (planned and actual production/progress and resource usage comparison), and replanning can be carried out at each level.

Exercise 2.6

A WBS is chosen to suit the needs of each project. It is an accepted way of representing how a project is broken into finer components. However, it is possible to not be systematic and describe the work in an unstructured list.

What is the difference between an unstructured activity list and a WBS? Why is the latter preferred?

Example: Programs

On major projects, the communication of planning outcomes in forms such as programs could be expected to occur on several levels. The levels roughly align with organizational levels. Senior staff may only be interested in broad issues. Mid-level staff would show interest in more detail. Individual production groups would work with still more detailed information. There is no consensus as to the number of levels used or what each level is called.

The programs at the higher levels are integrations of programs at lower levels. The lower levels are equivalent to subprojects, sub-subprojects, ... Resource, cost, and duration information at the lower levels is folded up to higher levels (Figure 2.7).

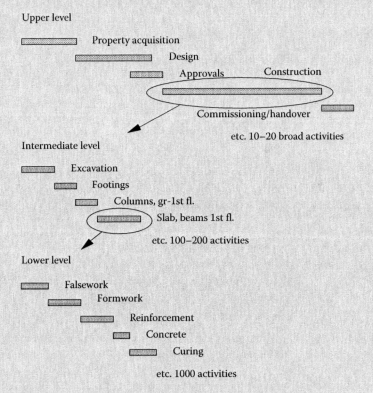

FIGURE 2.7
Different levels of programs, for example, construction of a multistory building. (Work parcel/activity names have been abbreviated because of space limitations—work items are implied.)

Exercise 2.7

Identify five hierarchies in your personal, sporting, social, or work life.

2.3.1 States, Controls, and Outputs

States, controls, and outputs exist at all levels of a hierarchical decomposition of a system.

Example: Structures

Controls that a designer might use at each level are

Structure level—structure stiffness
Member level—member stiffness
Element level—rigidity
Material level—material modulus

Behavior (state, output) at each level might be described by

Structure level—nodal displacements
Member level—nodal forces
Element level—curvature
Material level—stress

Example: Projects

For understanding projects, it is convenient to build up the fundamental relationships from the lower levels, referred to here (from the lowest level upward) as constituents, elements, and activities.

Constituents feed up to elements; elements combine to give an activity; and activities combine to give a project (and projects can be combined to give a "program").

Controls that a planner might use at each level are

Project level—work methods, overlapping activity relationships, concurrent/parallel activities, expediting materials, and project duration
Activity level—quantity of resources, distribution in time of resources, resource usage, compression/lengthening of activities, splitting activities, available float, and work calendars
Element level—number (quantity) and type of each resource
Constituent level—resource production rates

Exercise 2.8

Projects have a well-defined hierarchy: project–subproject–activity–element–constituent. Why do practitioners not take more advantage of this hierarchy in their work; instead they often tend to mix calculations on all levels?

2.3.2 Single Level

The development of multilevel systems theory has only taken place in the last several decades and is ongoing. Useable results are now starting to emerge. Historically, attention has been directed at the far simpler single-level case with a view to the internal composition of the system and the relationship of this composition to the levels of a multilevel system. It still remains convenient, both conceptually and computationally in many cases, to integrate the subsystems and work with a single level.

Exercise 2.9

It is sometimes said that a capitalist-oriented government acknowledges and encourages hierarchies in society, while a socialist government tries to eliminate hierarchies. What levels in society do you observe?

In nature, there is a food chain starting at the microscopic level and progressing with animal size:

Plant ⟶ Herbivore ⟶ Carnivore

or

Are these examples of hierarchical multilevel systems?
Does our system of government

Local ⟶ State ⟶ Federal

represent a hierarchical multilevel system?
Give other examples that you have observed of hierarchies in nature.

Three Men on Class (BBC-1, 1966); published in *No More Curried Eggs for Me* (Wilmut, 1982).

The sketch presents a humorous view of hierarchies. It involves three comedians, John Cleese, and the "Two Ronnies"—Barker and Corbett. The three stand in a line facing the camera and give different viewpoints on different subjects. With different heights of each actor, viewpoints are given in turn, from tall (Cleese) to medium (Barker) to short (Corbett), on subjects of class, and breeding, and ending with superiority:

Cleese: "I get a feeling of superiority over them."
Barker: "I get a feeling of inferiority from him (Cleese), but a feeling
        of superiority over him (Corbett)."
Corbett: "I get a pain in the back of my neck."

2.4 Staged Systems

Staged systems imply systems connected in series. One system comes after another (for example, Figure 2.8). The terms "phase" and "stage" might be used interchangeably. Examples include serial queuing systems (Figure 2.9), phases or stages in a project, and linear or sequential projects (Figure 2.10).

2.4.1 Project Phases

Based on observation, projects are seen to be separable into phases. Generally, the phases will occupy different time spans. Different writers give the phases different names, but the intention is the same. The naming changes with the

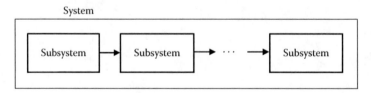

FIGURE 2.8
Staged system representation.

FIGURE 2.9
Serial queue. Customers denoted with circles.

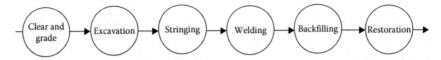

FIGURE 2.10
Linear project (for example, pipelaying) representation. (Activity names have been abbreviated because of space limitations—work items are implied.)

type of project. There is no definitive naming of phases, or number of phases. Some texts consider four generic phases:

1. Initiation or concept
2. Development
3. Implementation or execution
4. Termination or completion

Some writers also choose to include an additional phase, an "asset management" phase, at the end.

The choice of what constitutes a phase may depend on such matters as a time division, budget division, or resource-type usage.

Exercise 2.10

Do you think it would assist the advancement of project management practice if users could agree on a fixed number of phases and naming for all project types?

Should projects be forced into a rigid phasing mould, or should the mould be changed to suit the project?

Divisions between phases may not necessarily be characterized by noticeable discontinuities. Trouble with using the idea of phases comes about when it is recognized that there can be a blurring of activities across phase boundaries. All project activities are not sequential or serial in nature; feedback occurs between activities as ideas and information are consolidated.

The idea of phases, nevertheless, provides a useful way of managing projects.

Considered as a system, a project can be broken into subsystems with interaction between the subsystems. A project then can be thought of as being composed of interacting subprojects called phases (Figure 2.11).

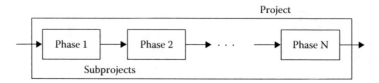

FIGURE 2.11
Project phases as subprojects.

Exercise 2.11

Phases are not totally separable. There is always some overlap between phases, no matter how chosen. Is then the notion of phasing a suitable way forward in trying to understand project management?

Exercise 2.12

Could the management of some projects be hindered through the promotion of the discontinuity that phasing introduces?

Exercise 2.13

Phases may not be totally separable; they may be strongly interdependent with overlapping. Fast-tracked projects have overlapping phases; a later project phase may be started prior to the completion and approval of all the previous phases. Is the notion of phasing still effective under such circumstances?

Example: Moviemaking

Moviemaking effort can be divided into phases where each phase represents the completion of activities/tasks, or relates to decision points, particularly whether or not to proceed to the next phase.

A possible movie project phasing terminology is

1. Initiation
2. Preproduction
3. Production
4. Postproduction

Example: Pharmaceutical Product Development

Phasing for new pharmaceutical product development, conducted within any statutory constraints, may take the form of

1. Preclinical testing: selection of sample and test regime
2. Clinical testing: selection of sample and test regime
3. Registering product and patenting product and name: government approval
4. Production and marketing

Example: Software Development

A phase description for a computer software project might be

1. Project definition
2. Business specification
3. Design
4. Code and test
5. User acceptance test
6. Documentation
7. Implementation

Example: Chemical Process Plant Construction

The following gives a breakdown of activities for a chemical process plant project. The activities have been divided into pre-investment and investment phases.

1. Pre-investment phase
    a. Project scope
    b. Project development
2. Investment phase
    c. Project implementation: owner and contractor
    d. Project initiation
    e. Process engineering
    f. Project engineering

Example: Biological Life Cycles

1. Germination
2. Growth
3. Maturity
4. Death

Example: Team Building

A phasing with rhyming appeal used in team building is

1. Forming
2. Storming
3. Norming
4. Performing
5. Adjourning

Exercise 2.14

What commonalities regarding phasing did the earlier example project types have?
  What can be learned by looking at projects from different industries?

Example: Humorous Phasing

Some humorous project phasing that may be seen on office notice boards is

1. Enthusiasm
2. Disenchantment
3. Panic
4. Search for the guilty
5. Punishment of the innocent
6. Decoration of the nonparticipants
7. Enrichment of the legal profession

Some related cynical musings include the fiction (*ASCE, Journal of Management in Engineering*, Vol. 9, No. 4, pp. 303–304):

Phase 1—fear: A frazzled engineer slumps in his chair, stares glumly at his profit/loss statement, chomps antacid tablets like candy, and worries about where his next job is coming from.

Phase 2—self-delusion: Rejecting harsh reality, the engineer busies himself with the foolhardy pursuit of illusory projects. Phone calls are made, letters of interest are mailed, and false flattery is lavishly doled out. The irrational hope of this phase can be likened to placing a written message in a bottle, throwing it into the Atlantic Ocean, and expecting to hear back from your cousin in England.

Phase 3—prevarication: A marketing offensive is launched with zealous fervor, complete disregard for the facts, and a proposed budget just slightly larger than the profit that could ever be realized from the project being pursued.

Phase 4—glory: Through some strange twist of fate, the project is awarded, backs are slapped, champagne corks are popped, and all is right with the universe.

Phase 5—boredom: Excitement turns to tedium as the staff wallow like overfed sows in an overabundance of chargeable man-hours and a deadline in the distant and unforeseeable future.

Phase 6—confusion: Where did the file go? Who's the project manager? What was it we were supposed to be designing again?

Phase 7—panic!: The budget is exhausted, the schedule is blown, and the major technical issues are unresolved. The stage is set for …

Phase 8—finger pointing: A tidal wave of creative energy is unleashed as an imaginatively conceived assortment of vendors, subconsultants, government officials, bad weather, illnesses, and various acts of God are trotted out as scapegoats responsible for the unfortunate status of the project. An engineer's years of schooling are put to full use as this elaborate version of my "Dog Peed on My Homework Assignment" is developed.

Phase 9—wonder: A wide-eyed engineer is overwhelmed with childlike amazement as he gazes at the unbelievable result of this irrational, painful, and morally reprehensible process—a successfully completed engineering project.

Phase 10—fear: A frazzled engineer slumps in his chair, stares glumly at his profit/loss statement, chomps antacid tablets like candy, and worries about where his next job is coming from.

State representation: Figure 2.12a shows a project broken into stages based on periods. Each period may be any chosen duration and can differ from other periods. Figure 2.12b shows the (planned) baseline for the case of a single state/output $x(k)$, $k = 0, 1,..., N$. Such a representation is only suitable for the single state/output case. For the case of multiple states/outputs, a state space representation such as Figure 2.12c is necessary. Figure 2.12c shows the

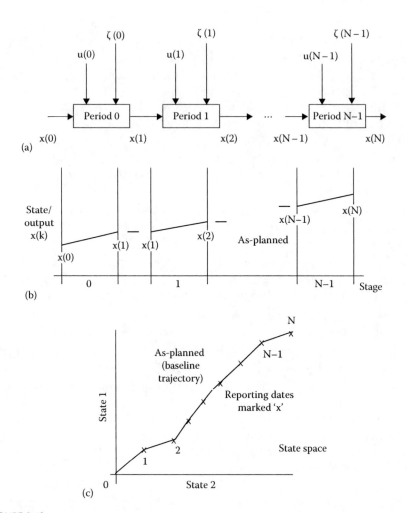

FIGURE 2.12
(a) Project periods/stages. (b) (Planned) baseline for single state/output. (c) (Planned) baseline for multiple (here 2) states/outputs.

(planned) baseline for the example situation of two states/outputs, but it can be extended to higher dimensional state spaces for multiple states/outputs. State and state space are discussed later under state equation models.

2.5 Model Development

Model development goes through steps.

Firstly, significant entities and relations in the system are selected. A model is chosen and examined or experimented with. Various deductions may

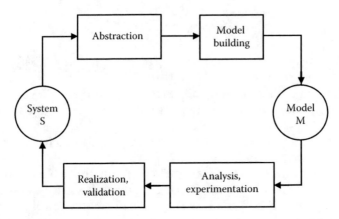

FIGURE 2.13
Model development process.

be drawn. These are then transferred to the original system (realization) where validation is carried out. The validation establishes whether the model is a suitable representation of the system or not. Where the suitability is less than required, the model development process starts again or refines the original proposal. By a series of iterations, a suitable model may be found (Figure 2.13).

Exercise 2.15

Assume that you wished to understand the motivational behavior of a person. What type of models might you consider? How would you go about refining these models?

Exercise 2.16

The excuse is often used that people can be unpredictable and that is why they cannot be modeled. But does not this just reflect that the type of model trying to be used is not appropriate? For example, unpredictability can be incorporated by using probabilities, such that input, state, and output variables become probabilistic. Or does this make the model unnecessarily complicated?

Monty Python's Flying Circus (BBC Television, UK, 1969–1974)
  Frontiers in Medicine
  Model development practices are disclosed in this humorous Monty Python sketch.

The sketch outlines a study of penguins performed in an attempt to understand the human mind, among side references and distractions to tennis.

The starting theory is that the penguin is intrinsically more intelligent than the human being.

On a diagram showing a man and a penguin, the penguin's brain is smaller. However, on drawing a penguin the same size as a man, "the penguin's brain is still smaller. But, and this is the point, it is larger than it was. For a penguin to have the same size of brain as a man the penguin would have to be over sixty-six feet high.

"This theory has become known as the waste of time theory and was abandoned in 1956."

On carrying out IQ tests, "the penguins scored badly … but better than BBC program planners. The BBC program planners' surprisingly high total here can be explained away as being within the ordinary limits of statistical error.

"These IQ tests were thought to contain an unfair cultural bias against the penguin. For example, it didn't take into account the penguins' extremely poor educational system. To devise a fairer system of test, a team of our researchers spent eighteen months in Antarctica living like penguins, and subsequently dying like penguins—only quicker—proving that the penguin is a clever little sod in his own environment."

Tests were then carried out on the penguins in their own environment, here a pool at the zoo. Penguins were asked mathematical questions. This removed the environmental barrier, but not the language barrier. "The penguins could not speak English and were therefore unable to give the answers. This [issue] was removed in the next series of experiments by asking the same questions to the penguins and to a random group of non-English-speaking humans in the same conditions [standing in the zoo pool].

"The results of these tests were most illuminating. The penguins' scores were consistently equal to those of the non-English-speaking group."

The sketch ends with penguins taking on human jobs, and BBC program planners being fed at the zoo.

2.5.1 Scientific Method

Disciplines such as engineering and the sciences have developed over many hundreds of years. They have established bodies of knowledge, developed largely through the so-called *scientific method*.

The scientific method is based on a process of observing behavior, postulating or hypothesizing a model or theory of behavior, and conducting experiments, observations, or measurements. Should the experimentation support the *hypothesis*, a *law*, *model*, or *theory* is established. Should the experimentation not support the hypothesis, the original hypothesis/model/theory is modified or replaced with another and the process repeats itself (Figure 2.14).

> The great tragedy of science—the slaying of a beautiful hypothesis by an ugly fact.
>
> **T. H. Huxley**

The essence of any sound and rigorous research in engineering or science is based on this scientific method. It is suggested that all disciplines adopt a similar approach.

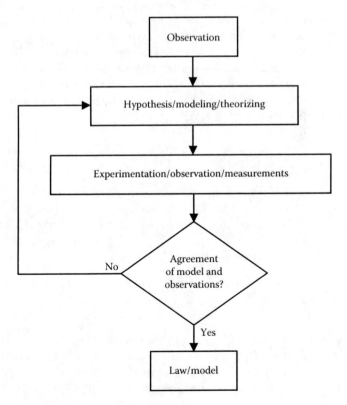

FIGURE 2.14
Essence of the "scientific method."

Hypotheses may come about through

- *Deduction*—Reasoning from the general to the specific
- *Induction*—Reasoned generalizations based on a number of specific instances

Deduction and induction work in opposite ways. An inductive generalization is much like a hypothesis and may be revised as new information comes to hand.

The terms *model, theory, hypothesis,* and *law* might be used interchangeably by people. More usually, hypothesis is something to be tested and demonstrated to hold or be refuted, while the others may or may not have been already demonstrated or refuted. Some people use the term "theory" (and even "model"), particularly in the popular management literature, implying that the proof is self-evident or not necessary. That is, the theory is never proven, but rather is held out to be a compelling truth, even though it may not be compelling or even a good theory.

In this book, the term "hypothesis" is used in the earlier sense, and not in any statistical sense. In the discipline of statistics, use is made of a pair of hypotheses—a "null hypothesis" and an "alternative hypothesis." The null hypothesis generally corresponds to a general or default position, for example, there is no relationship between two variables. It can generally not be proven to hold; it can only be rejected or fail to be rejected. The alternative hypothesis asserts that a particular relationship exists. The alternative hypothesis need not be the logical negation of the null hypothesis.

The systems configuration of investigation, covered later, aims to short-circuit the process of obtaining suitable models.

Darwin: Natural Selection

Charles Darwin in his "Origin of Species" (1859) used the term "natural selection" to explain his theory of evolution. Based on a multitude of observations, natural selection remains the primary explanation for adaptive evolution.

Principles of Sport

In competitive two-team games, such as football, the team with the ball (or equivalent) is in attack; the team without the ball is in defense. The attacking team tries to score goals/points; the defending team tries to stop the other team from scoring. The game goes to and fro with teams alternating between attack and defense. Through observation, it is possible to come up with the following principles or hypotheses (Carmichael).

Principle A: In attack, the return (in terms of goals/points scored) reflects the degree of uncertainty (in the defense's mind) associated with the play of the attacking team.

Principle D: In defense, the return (in terms of goals/points prevented) inversely reflects the degree of uncertainty (in the defense's mind) associated with the play of the attacking team.

That is, attacks try to maximize uncertainty, while defenses try to minimize uncertainty. Uncertainty is introduced as soon as an attacking play has more than one possible outcome, each with an associated probability of occurrence. Uncertainty is reduced by restricting the number of possible outcomes and making the outcome more predictable.

The principles can be readily demonstrated to be true by collecting field data on different plays that attacking teams use. For a sufficiently large data set, the attacking plays that lead to most goals/points show high degrees of uncertainty—that is, they are difficult for defenses to predict; and the attacking plays that lead to least goals/points show low degrees of uncertainty—that is, through poor attack or good defense, the number of attacking options are restricted. The principles can also be demonstrated to be true through the use of probability models.

All team coaches instill such thinking into their teams, though the coaches would almost surely not have heard of these principles. Some examples, in association football (soccer), are defensive plays that restrict uncertainty include structured formations, rehearsed rules, man-on-man marking, playing one extra defender (sweeper) than there are attackers, closing-down the player on the ball, and erecting a player-wall after a free kick has been awarded; attacking plays that increase uncertainty include unstructured formations, few rules, movement of players off-the-ball, having multiple attack possibilities (not doing the same thing with every attack), unmarked players arriving late in front of goal, and balls centered in front of goal.

The more skill, speed, peripheral vision, quick thinking ability, fitness, experience, and training a player has, the more scope that player has to reduce uncertainty while acting as a defender or increase uncertainty while acting as an attacker. Conversely, the less-well prepared a player is, the less likely that player will do something in attack that is difficult to predict, while the player is less likely to perform well in defense.

Ongoing throughout a game is the desire for each side to have the ball, and tactics to win back the ball, because only with the ball can a side attack.

Accordingly, the mechanics of two-team competitive sports are straightforward to model and coach. The people-handling side (both on- and off-field), however, is a far more difficult affair.

Exercise 2.17

The Hawthorne experiments. In the 1920s and 1930s, Elton Mayo conducted some experiments in factory-type surroundings, varying work-area lighting and observing its influence on production. The production of the group selected for the study, and the production of a comparison group, were found to increase independently of the lighting levels.

In further experiments, other conditions (salaries, coffee breaks, workday and workweek lengths, flexible breaks, etc.) were varied, and their influence on production observed. The production of the group selected for the study, and the production of a comparison group, were found to increase independently of the conditions.

The conclusion: the special attention given to the groups caused the groups to increase their efforts. This is known as the Hawthorne effect.

Design an experiment whereby the results would add support to the Hawthorne effect. Pay particular attention to how you might keep all variables constant between experiments except the one or two variables that you are interested in obtaining data on.

Exercise 2.18

Scientific management: Frederick Taylor (1856–1915) conducted many experiments related to worker production. He examined and timed workers' movements on jobs. With such time studies, he then broke down each job into its components and designed the quickest and best way to do work.

One conclusion from his work is that a team of people working together, each tending expertly to one or a few tasks, can outproduce the same number of people each performing all of the tasks. An assembly line is an example of this.

Design an experiment whereby the results would add support to Taylor's conclusion. Pay particular attention to how you might keep all variables constant between experiments except the one or two variables that you are interested in obtaining data on.

Exercise 2.19

Maslow's hierarchy of needs: Abraham Maslow developed a hierarchy of needs in terms of the following levels (highest to lowest):

> Self-fulfillment
> Ego
> Belonging
> Safety/security
> Physiological

As a lower level is satisfied, a person's needs shift to the next higher level. Lower levels no longer motivate behavior.

This hierarchy explains certain behavior in workers such as why one particular job may appeal over another, even though the pay might be less. Generally, it could be expected that the needs outlined in the lower levels are satisfied in many workers (in developed countries), and therefore the motivators lie in the upper levels. Upper-level needs are satisfied, for example, by praise, listening to the suggestions and complaints of the workers, allowing workers to use their own initiatives, and reducing the amount of supervision.

People with specialist skills or crafts can obtain self-fulfillment through their work. Though less obvious, people without skills or crafts, but who take pride in their work, can also obtain self-fulfillment. Seeing something complete, something tangible, and having made a contribution to this "thing" is also satisfying. People who do not practice a particular skill but who supervise and hence who do not contribute anything tangible may have more difficulty obtaining satisfaction.

Design an experiment whereby the results would add support to Maslow's hierarchy. Pay particular attention to how you might keep all variables constant between experiments except the one or two variables that you are interested in obtaining data on.

2.5.1.1 Nonrigorous and Management Models

Not all models that appear in the literature are established with rigor. By the scientific method, careful observation and equally careful inference or deduction from these observations are fundamental in establishing the exact nature of behavior. The nonrigorous approach tends to be opinion and anecdote based and lacks generality; this is compensated for with hype and buzzwords.

The field of popular management is very crude when compared, for example, with established engineering and science disciplines of age. Popular management models tend to be largely verbal models with all their attendant drawbacks, as well as being nonrigorous. Most popular management contributors are not able to model otherwise because of their educational backgrounds.

The primary goal of the scientific method is to investigate the true nature of what is around us, and the mechanisms of change from one form to another. It starts by observation. This is followed by formulating a hypothesis, theory, or model—a possible explanation of the facts, something that can predict the outcome of other situations that have not yet been subjected to widespread investigation, under a variety of different conditions. After many tests and revisions, the hypothesis may be accepted and become a law.

When you read any literature, critically appraise the basis of what is written. For example, when reading the popular management literature (or attending popular management seminars), or non-peer-reviewed articles, including those on the Internet, do not accept what is written (or said) without question. By all means, take any new ideas on board, but be aware of their origins.

It would seem reasonable (at least to an engineer or scientist) therefore that a good place to start the development of a body of knowledge in any discipline, for example, management, would be to use the "scientific method." This appears to have been the case in management studies in the first half of the last century. But many recent "contributions" to the popular management literature appear to have forgotten about rigor.

> While we see management as parallel to other professions, we recognize that it is still in its infancy. There are thus lessons to be learned by management in the evolution of other professions that sought to develop clear language and to act on the basis of high ethical standards, logic, and proven ideas.
>
> **Hilmer and Donaldson, 1996**

Exercise 2.20

An equity model. Equity is defined as

$$\text{Equity} = \frac{\text{Input}}{\text{Outcome}}$$

that is, the ratio of the input of the individual to the outcome or rewards resulting from that input.

How an individual perceives the equity in a work task, relative to another individual's work, establishes the degree of motivation, performance, and satisfaction in the work.

Typically, people compare pay received and effort and skills expended with others in similar jobs, both within an organization and between organizations. A person may perceive an imbalance relative to another person in either a positive or negative sense. Where an inequity is perceived, a person may adjust either the numerator or denominator in the equity expression up or down as the situation warrants.

Alternative behavior includes the following:

- A person may rationalize the situation by altering his/her perceptions of someone else's input or outcomes.
- A person, rather than adjusting his/her numerator or denominator in the equity expression, may coerce or otherwise get the other person to adjust his/her numerator or denominator.

Equity refers to perceptions when a comparison is made with fellow employees, in particular the perception that fellow employees receive a comparable reward for a comparable input.

How would you go about demonstrating the validity or otherwise of this equity model? Could you set up an experiment?

There have been a number of well-publicized events involving large high-profile companies:

1. The emotional significance of interclass pay equity was shown in the angry messages that were posted on a company's internal computer bulletin board when a chief executive officer's record compensation was announced at the same time that the profit-sharing formula was revised to be less generous to other employees.
2. A similar situation occurred when another company negotiated wage concessions from its unionized employees and then announced that executives would receive large bonuses. The employee outrage, that ensued, led the company to cancel the bonuses.

Do these events confirm the equity model? Or are they just persuasive support?

Inputs (for example, skills) are typically more ambiguous than outcomes (for example, pay). Furthermore, inputs are subject to strong self-enhancing perceptual biases that cause people to give themselves more credit than is deserved, and others less. This bias in social comparisons of work performance can be demonstrated by studies in which typically at least two-thirds of employees rate their performance as being in the top quartile of those with similar jobs. Moreover, the same self-enhancing bias appears when people compare their group with other groups. Thus, lower-level employees' perception of pay equity will be primarily determined by pay differentials between lower and higher organizational strata, and only to a lesser extent by input differentials. Under such circumstances, how can the equity model be confirmed?

Exercise 2.21

Modeling people and organizations: The modeling of people behavior and organizations is considered crude compared to that of mechanical systems. And the only way our understanding of people is going to improve is to go from the present imprecise verbal models to quantitative and mathematical models of human and organizational behavior. Some people counter this by saying that people are so complicated that they will defy any quantitative modeling. What do you think models of people may end up looking like, or will we be forever in the dark?

2.5.2 False Causality

Be aware of the distinction between causality and correlation. Two variables might be highly correlated, yet there may be no causal relation between the variables. Be aware when doing regression and correlation studies that the statistics do not address causal relations.

A true engineer or scientist will always look for the underlying truth as given in causal relations, rather than a more expedient statistical study of correlation. The causal relation evolves from thinking about the system, while correlation is obtained by processing data.

Beware of introducing false causality in modeling. The following examples show how, based on data alone, there might be thought to be a causal relation.

Helmets Cause Head Injuries (Source Unknown)

During World War I, the cloth caps worn by soldiers were eventually replaced with metal helmets. However, head injuries increased after this change. Conclusion: Helmets cause head injuries. But do they?

Answer: The reason that head injuries increased after the introduction of helmets is because before the helmets were introduced, people who got hit in the head generally died. After the helmets came along, there were fewer deaths but more head injuries.

Lactic Acid and Muscle Fatigue

For many years, it was believed (based on electrical impulse experiments on frog's muscles) that lactic acid build up in muscles during exercise causes fatigue. This is "guilt by association." There is correlation between lactic acid levels in muscles and fatigue levels, but does lactic acid cause fatigue? There is an alternative view that lactic acid is necessary for muscle performance.

Coffee and Miscarriages

Some medical research showed a positive correlation between the coffee intake of pregnant women and miscarriages. Women who had larger intakes of coffee were more inclined to have miscarriages. Conclusion: Coffee causes miscarriages.

However, other researchers point out that healthy women have morning sickness and hence are not in the mood to drink coffee, whereas unhealthy women are less inclined to morning sickness and hence are able to drink coffee. That is, the causality is between the (poor) health of the women and miscarriages, and coffee has nothing to do with it.

The research is ongoing.

Monty Python's The Holy Grail (Columbia Pictures, UK, 1975)
How to tell a witch...

This Monty Python sketch covers a number of humorous attempts to establish causality and hence prove that a woman is a witch, thereby allowing the villagers to burn her.

To prove that the woman is a witch, the villagers first dress the woman all in black with a carrot tied around her face on top of her nose and a black paper hat on her head. And she talks strangely because her nose is closed by the carrot. This ruse was quickly uncovered.

Next, one villager claims that the woman turned him into a newt. This, no one believed.

Next, it is argued that wood burns, and so if witches get burned, then witches are made of wood. And to tell whether she is made of wood, the logic is to throw her into the pond. Wood floats, and so if she floats, then she is made of wood. But ducks also float, and so if she weighs the same as a duck, then she is made of wood.

The sketch ends with the weighing of the woman, who is found to weigh the same as a duck. The woman concedes to this logic.

Exercise 2.22

One well-known university promotes one of its "prestigious" postgraduate degrees by quoting survey data of students' salaries before enrolment and students' salaries after graduation, and noting that the difference between these two salaries exceeds the cost of the degree. The university imputes a relationship between having the degree and the salary that the graduate earns. What is wrong with drawing such a causal relationship? Is the graduate's salary due to having the degree, or would the individuals have achieved such salaries independently of the degree, or is there some other explanation for the graduates' salaries (such as in the time between enrolling and graduating, career promotion, or change had occurred)?

Some good advice: *Use a thicker pen when trying to match a linear relationship to a set of data.*

2.5.3 Adages, Truisms

A number of adages and truisms exist on modeling. These commonly go by the names of "laws," "principles," or "rules." These can be likened to hypotheses, whereby their proofs are left to the observer to confirm or refute through everyday living.

Example: Murphy's Law

The origin of Murphy's Law, the original principle *If anything can go wrong it will* is unclear. Some of the published attribution is not convincing. Possibly, the naming came about because Murphy is a distinctly Irish name, and the Irish are known for their quirky humor. The book Murphy's Law, by A. Bloch, Methuen, London, 1986, lists numerous extensions to Murphy's Law, including

Nagler's Comment on the Origin of Murphy's Law
Murphy's Law was not propounded by Murphy, but by another man of the same name.

Reliability Principle
The difference between the Laws of Nature and Murphy's Law is that with the Laws of Nature, you can count on things screwing up the same way every time.

O'Toole's Commentary on Murphy's Law
Murphy was an optimist.

Goldberg's Commentary
O'Toole was an optimist.

There are many extensions and corollaries of the basic principle behind Murphy's Law, and again the authorship of these is unclear. All give a humorous view of laws guiding life. Murphy has something to say on most matters in life.

Example: Pareto Rule/Principle

The Pareto Rule/Principle is named after Vilfredo Pareto (nineteenth century). It is sometimes called the 80:20 Rule/Principle. Originally applied to wealth distribution, it has since been generalized: 80% of the outcomes (outputs) are the result of 20% of the influences (inputs). The numbers 80 and 20 are not to be interpreted precisely.

There are many applications in engineering. For example, 80% of a cost estimate is the result of 20% of the component cost items; 80% of a person's output is due to 20% of that person's tasks.

Example: Parkinson's Law

Parkinson's (first) law (Parkinson, 1957)
Work expands so as to fill the time available for its completion. The thing to be done swells in perceived importance and complexity in a direct ratio with the time period to be spent in its completion.

Parkinson's (second) law
Expenditures rise to meet income.

Parkinson's law of delay
Delay is the deadliest form of denial.

Mrs Parkinson's law (Parkinson, 1968)
Heat produced by domestic pressure expands to fill the mind available from which it can pass only to a cooler mind.

2.6 Classification

There are a number of ways that models can be classified. For example, models may be thought of as

- Mathematical models (for example, a mathematical equation)
- Verbal models (for example, words or a sentence)

or

- Material or physical models (for example, a scale)
- Symbolic or formal models
- Graphical models

or

- Concrete (empirical) models
- Conceptual models
- Formal models

Each model type has its strengths and weaknesses, and area of best applicability.

Another view classifies the model *according to the type of system being studied*:

- Static/dynamic
- Deterministic/stochastic

Alternatively, the *method or working principle* of the model may be the basis of the classification, for example,

- Scale
- Analogue
- Structural
- Mathematical
- Abstract

Or the *model's function* may be used, for example,

- Exploratory
- Heuristic
- Descriptive
- Reductive
- Explanatory
- Operationalizing
- Formalizing

One model may fit more than one of these classifications.

Example: Mathematical Models

Any mathematical equation, whether algebraic, differential, matrix, or other, can be regarded as a mathematical model. To progress thinking in any discipline, mathematical models are favored for their definiteness. They permit calculations to be made. Mathematical models are favored in engineering and science.

Example: Concrete (Empirical) Models

Examples of these model types are the planetarium (a model of the solar system) and the ruler (a model of real numbers). Many studies in coastal hydraulics and structural engineering laboratories use empirical models.

Example: Conceptual Models

Conceptual models tend to be more commonly used than concrete models. Examples include networks, maps, figures, patterns, drawings, and the periodic table of the elements.

Designers frequently use conceptual models in the form of technical drawings. They are models of the designers' thoughts and images, which may be regarded as conceptual systems.

Example: Formal Models

Formal models are typically symbols or abstract language. Examples of formal systems/formal models include mathematical systems and formulae. Formal models are commonly used for all system types.

An example is the use of a differential equation to describe dynamic motion. Other examples are the mathematical models used in economics.

Example: Verbal Models

Much of the knowledge in management and the social sciences is contained in words or verbal models. Such models lack definiteness because of the imprecision of language. Verbal models might be considered the first step in the evolutionary development of modeling in any discipline, but largely prevent further development of thinking unless a transfer to something quantitative is made.

Example: Physical Models

Architects commonly have small-scale representations of buildings made out of cardboard, foam, plastic, balsa wood, and similar materials in order to convey to their clients what the full-scale buildings may look like. Equivalently, three-dimensional computer graphics packages may be used to portray the same information.

Example: Mathematical Models—Learning Curves

Generally, the more times a task is repeated, the better and quicker people (and groups and organizations) become at doing that task (until boredom or alienation occurs). Knowledge of this is used in planning work. Figure 2.15 shows this schematically in terms of units or tasks performed and is called a learning curve.

The learning curve is commonly described by an equation of the form

$$T_n = T_1 n^b$$

where
   T_n is the average work/unit after n units
   T_1 is the work/unit for the first unit
   b is an index of learning (b < 0)

Different values of b, the index of learning, represent different values of the rate of learning. Both T_1 and b require estimating. Typically, these are adapted from similar situations or historical records.

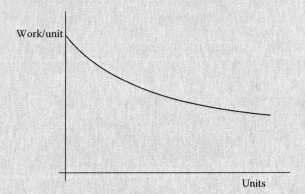

FIGURE 2.15
Learning curve.

Example: Graphical Model—Project Programs

Programs used on projects may be any of the following:

- (Connected) bar (Gantt) chart
- Time-scaled network diagram
- Cumulative production plot
- Activity date/time listings/timetable
- Marked-up drawings

They ideally derive from a critical path analysis performed on the project network. They may commonly be presented

- Over different project levels
- Over different time horizons

Associated with such programs, there may be a need to also communicate resource and money requirements in the form of

- Resource plots: S (resource) curves
- Budget, cash flow: S (money) curve

Example: Graphical Model—Multiple Activity Charts

A multiple (multi-) activity chart shows resource (people and/or equipment) usage to a common time scale, much like bar charts. Separate bars are drawn for each resource and show periods of idleness and periods when the resource types are utilized. See Figure 2.16. Times for such charts are typically based on field data, and hence multiple activity charts are more commonly used to represent existing work practices.

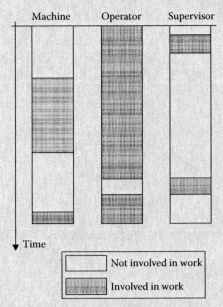

FIGURE 2.16
Example multiple activity chart.

Example: Graphical Model—Process Charts

Process charts portray a sequence of activities diagrammatically by means of a set of process chart symbols to help a person visualize a process as a means to examining and improving it. See Currie (1959) and ILO (1969).

An *outline process chart* gives an overall picture by recording in sequence only the main operations and inspections.

A *flow process chart* sets out the sequence of the flow of a product or a procedure by recording all activities under review using appropriate process chart symbols. A *resource type flow process chart* records how the resource is used.

Process charts are constructed using standard symbols defined in Figure 2.17. The work is broken down into its component activities, which are represented schematically by these symbols. Further breaking down of the work may be possible and is sometimes done.

An *outline process chart* is useful in design and planning. A method study is carried out before any work begins. The chart gives an overall view of the work and graphically shows the sequence of activities where inspections occur and where extraneous materials, information, etc., are introduced. It does not show who carries out the work or where the work is carried out. Accordingly, only the operation and inspection symbols are used.

Figure 2.18 shows an example outline process chart. Activities can be numbered in order, and descriptions are placed adjacent to the symbols. Activity durations can also be placed on the chart. Variants on the theme in Figure 2.18 occur, for example, when work is divided or reprocessed, or alternative means are possible to perform an activity; such variants lead to branching, loops, and multiple parallel paths, respectively.

Symbol	Activity	Predominant Result
$\bigcirc$	Operation	Produces, accomplishes, furthers the process
$\Rightarrow$	Transport	Travels
∇	Storage	Holds, keeps, or retains
D	Delay	Interferes or delays
$\square$	Inspection	Verifies quantity and/or quality

FIGURE 2.17
Symbols for process chart construction. (From Currie, R. M., *Work Study*, Pitman, London, 1960; International Labour Office (ILO), *Introduction to Work Study*, ILO, Geneva, Switzerland, 1969.)

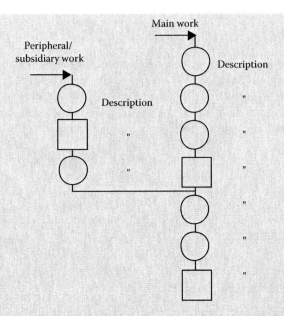

FIGURE 2.18
Example outline process chart.

A *flow process chart* extends the outline process chart to include transport, delay, and storage activities. The chart may be drawn from the point of view of the worker or the equipment or material used by the worker. It is similar in appearance to the outline process chart. Travel distances can be included to scale or by annotation.

Figure 2.19 shows an example flow process chart.

Activity	Oper'n	Trans.	Stor.	Delay	Insp.	Remarks
Wait						
Load						
Check						
Maneuver						
Travel						
Wait						
Unload						
Check						
Stockpile						
Total	2	2	1	2	2	

FIGURE 2.19
Example flow process chart.

Example: Planning

Formal planning includes the use of the following:

- Hierarchical models: The models are developed from element to activity to subproject to project.
- Multistage models: The project is broken down into stages, or equivalently the project duration is discretized.
- Probabilistic models: Activity durations and costs are described in terms of random variables.

Example: Depreciation Models

A number of models can be adopted for depreciating the value of assets, though the method chosen may be more determined by taxation requirements than by other reasons.

The models may be based on a *time basis* or a *usage basis*.

Time basis: Where the asset's declining value is considered to be dependent primarily on age, the asset's value is apportioned over its estimated life.

Different types of equipment are considered to have different life spans. For example, personal computers and measuring instruments might be assumed to have a useful life of 3 years, while expensive industrial machinery might be assumed to have a useful life of 5 years or more.

Common depreciation models might be described by the following method used:

- Straight line
- Reducing balance (declining balance, diminishing value)
- Sinking fund
- Sum of digits (sum of years, sum of year digits)

The last two models appear to be rarely used. The first two models appear acceptable to taxation authorities.

The straight line depreciation method is popular because of its simplicity.

Usage basis: On a usage basis, an asset is depreciated by the usage it gets in any one year compared to its estimated lifetime capacity for usage.

Free depreciation: An alternative to time-based and usage-based depreciation is to depreciate the value of an asset by its total value at the time of purchase. This is an extreme form of depreciation, but one that might be contemplated for assets with very short lifetimes. It may also be used, for example, where an asset is to be charged totally to a single project; for future projects, the asset represents no cost.

In practice, depreciation tends to be calculated in the simplest way possible. This is possibly a hangover from the pre-calculator days. But whichever method is used, the calculations should reflect as closely as possible the true declining value of the asset, taking into account its age and usage.

All models require the following information, which may only be best estimates:

- Original (purchase, new) value
- The economic life (time between obtaining the asset and disposing of the asset), or the recovery period for taxation purposes
- The residual (salvage, disposal, trade-in) value: commonly zero for taxation purposes

Straight-line depreciation: Straight-line depreciation reduces the value of equipment by equal annual amounts. For example, a piece of equipment costing $50,000, if it is assumed to have a life of 10 years and a residual value of $5,000 at the end of 10 years, will have an appreciable value (total depreciation) of $50,000 − $5,000 = $45,000, and an annual depreciation of $45,000/10 = $4,500. See Figure 2.20.

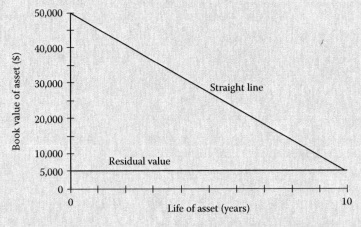

FIGURE 2.20
Plot of straight-line depreciation example.

The straight-line method is the simplest method to use and, possibly because of this, is widely used.

Let
  P be the initial value of the asset
  L be the residual value
  n be the number of years of expected life
  j be the year counter, j = 1, 2, ..., n

Then,

Rate of depreciation $= \dfrac{1}{n}$

Annual depreciation charge $= \dfrac{P-L}{n}$

Total depreciation at end of jth year $= \dfrac{(P-L)}{n} j$

Book value at end of jth year $= P - \dfrac{(P-L)j}{n}$

Diminishing value method: Also called declining or reducing balance depreciation, the depreciation is calculated as an annual fixed percentage.

Let
  P be the initial value of the asset
  L be the residual value
  n be the number of years of expected life
  r be the percentage depreciation per annum
  (expressed as a number less than 1)

Then,

$$\text{Depreciation at end of first year} = Pr$$

$$\text{Depreciation at end of second year} = (P-Pr)r$$

$$= Pr(1-r)$$

$$\text{Depreciation at end of nth year} = Pr(1-r)^{n-1}$$

$$\text{Total depreciation at end of nth year} = Pr + Pr(1-r) + \cdots + Pr(1-r)^{n-1}$$

$$\text{Book value at end of the first year} = P - Pr$$

$$= P(1-r)$$

Book value at end of the second year $= P - \{Pr + Pr(1-r)\}$

$$= P(1 - 2r + r^2)$$

$$= P(1-r)^2$$

Book value at end of nth year $= P - \{Pr + Pr(1-r) + \cdots + Pr(1-r)^{n-1}\}$

$$= P(1-r)^n$$

That is,

$$L = P(1-r)^n$$

or

$$r = \left(1 - \sqrt[n]{\frac{L}{P}}\right)$$

For the example of the piece of equipment costing \$50,000 with residual value of \$5,000 after 10 years,

$$r = \left(1 - \sqrt[10]{\frac{5,000}{50,000}}\right) = 0.206 = 20.6\%$$

Table 2.1 shows the declining value of the piece of equipment.

TABLE 2.1

Declining Balance Depreciation

End of Year	Depreciation (%)	Depreciation for Year ($)	Book Value ($)
0	20.6	0	50,000
1	20.6	10,284	39,716
2	20.6	8,169	31,547
3	20.6	6,488	25,059
4	20.6	5,154	19,905
5	20.6	4,094	15,811
6	20.6	3,252	12,559
7	20.6	2,583	9,976
8	20.6	2,052	7,924
9	20.6	1,630	6,294
10	20.6	1,294	5,000

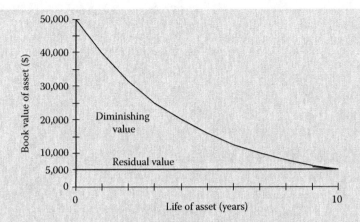

FIGURE 2.21
Plot of declining balance depreciation example.

See Figure 2.21.

The declining value method gives higher depreciation during the earlier years than the straight-line method. It is believed to represent the true situation more accurately than the straight-line method.

The previous calculation establishes an interest rate, which will reduce the value of the asset to its residual value at the end of its estimated life.

An alternative is that a taxation authority specifies, or a company decides, depreciation is to be at a given annual percentage.

Exercise 2.23

Consider the declining value of one of your assets, for example, a motor vehicle or computer. Estimate its initial purchase price, its lifetime, and any residual value at the end of this lifetime. Plot approximately how you believe the value of this asset will decrease with time.

Exercise 2.24

Consider a piece of equipment purchased for $100,000 with a useful life of 5 years after which its scrap, salvage, or residual value is $10,000. Calculate the depreciation using the straight-line method and the diminishing value method. Plot the results in a similar fashion to Figure 2.20.

Exercise 2.25

Give examples of each of the following model types:

- Scale
- Analogue
- Structural
- Mathematical
- Abstract

Exercise 2.26

Computer simulations are used to train pilots, plant operators, etc., as well as for entertainment in the form of games. What type of models are these? Or is such simulation a combination of a model and analysis using that model? [The analysis configuration is one where the system input and model are given, and the system output is derived.]

Exercise 2.27

Consider the model

$$Y = 2X$$

where
   Y is the number of goats
   X is the number of sheep

This would be called a mathematical model.
   Are the following verbal models or mathematical models?

(i)   The number of goats = Two × The number of sheep.
(ii) The number of goats equals twice the number of sheep.
(iii) The ratio of goats to sheep is two.
(iv) There are twice as many goats as sheep.
(v)   The number of goats is double that of the sheep.

Each says the same thing as the earlier mathematical model but are (mainly) in words.

Exercise 2.28

Recent attention in sport is to elevate the thinking behind the way sport is played. This not only includes diet, physiology, and training, but (and this is the subject of this exercise) understanding and devising tactics of play. This exercise is looking at the modeling of sport behavior. You can choose the sport.

What types of models do you think would suit sport the best? Should they be quantitative, mathematical, verbal, simulation-style, statistical, or other models? Carry out a brief search of the literature (for example, look at five journal papers or books) and see what types of models people are using to model various aspects of sport. Comment on what you see are the good points and the deficiencies of these models.

Exercise 2.29

The only way our understanding of people is going to improve is to go from the present imprecise verbal models to quantitative and mathematical models of human and organizational behavior. This exercise is looking at the quantitative and mathematical modeling of human and organizational behavior.

Some people say that human behavior is too complicated to model, but these people seem to be those without modeling skills. Other people say that proper modeling is necessary, but that we may have to rethink the types of models and develop new and different modeling to that used in the sciences and engineering to date.

What types of models do you think would represent people behavior the best? Should they be quantitative, mathematical, verbal, simulation-style, statistical, or other models? Carry out a brief search of the rigorous literature (for example, look at five journal papers or books) and see what types of models are being used to model various aspects of people and organizational behavior. Comment on what you see are the good points and the deficiencies of these models.

2.7 System Model Terminology

It is found convenient when describing systems to use some standard terminology for things that change (variables). Mathematical models are developed in terms of symbols for these variables and the behavior (including dynamic behavior—behavior over time) of the system studied.

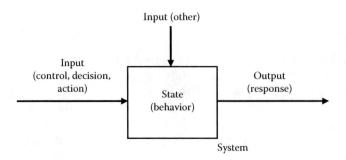

FIGURE 2.22
System representation.

Figure 2.22 gives a schematic representation of a system.

Arrows, although conventionally used to denote flows of the relevant quantities, are used in Figure 2.22 to imply dependence relationships. They facilitate the description of a system as sets of input–output pairs and are convenient when generalizing systems concepts (which have primarily been developed for flow systems) to nonflow systems (for example, infrastructure).

Input variables are of two types:

1. *Control*, decision, action, or design *variables* that can be manipulated or chosen by the engineer in order to drive the system behavior in a desired way, or according to some objective. The term "control" is generally preferred in this book, though some existing literature has already adopted the terms "decision," "action," and "design."
2. Those over which the engineer has no influence or chooses to have no influence.

Commonly, there will also be a *disturbance* (or noise or corruption), where variability or uncertainty is present. The disturbance prevents the hoped-for output being obtained or measured exactly, or the state being known exactly. The term "disturbance" is preferred in this book, though control system texts may use the term "noise." In projects, for example, the disturbance comes about because of a changing project environment brought about through delays, disruptions, adverse climatic effects, and so on. A disturbance might be portrayed as an input when a system is represented diagrammatically.

Output variables are variables reflecting the external, measurable, or observable behavior of the system. They are a consequence of the inputs and the system. Alternative terminologies are "response" and "performance."

State variables are variables reflecting the internal behavior of the system. They cannot be observed. They are a consequence of the particular values of the input variables, and the input variable–system behavior relationship. The state is usually assumed in models to contain all relevant historical information about the system; that is, the future state depends on the current state (plus future inputs) and not past states.

Where there is more than one state variable, each state variable can be plotted on an axis giving what is called a *state space*. For n state variables, the state space is n-dimensional. Figure 2.12c is an example state space.

Since state, control, and output each contain multiple descriptors, they can be conveniently represented as vector quantities. For example, if the state of a system is represented by the pair—position and velocity—then the state space can be thought of as being two-dimensional, and the state represented by a 2 × 1 vector.

Constants of the system are referred to as *parameters*. For simplification in modeling, variables in a more general model may be treated as parameters in a simpler model.

Example

Consider a framed structure.

State variables contain all the information regarding the internal behavior or state of the system. State variables may be typically bending moments, shear forces, and axial forces in the frame's members, deflections, etc., or stresses and strains at lower subsystem levels. All these are behavior variables.

The output or response indicates the level of outward or observable behavior (structure deflections, …) of the system.

The control/decision variables are those entities under the influence of the designer. They contain the information relating to the material properties, geometry, and physical properties (flexural stiffness, axial stiffness, …) of the various components of the system. They exert the control on the behavior of the system and may be freely chosen (manipulated) by the designer. In this sense, they are "input" into the model by the engineer.

Output variables may or may not be identified directly with state variables. Generally, the output, being a relevant measure of the state, will comprise part of the state description.

In some cases, the output and state may obey a one-to-one transformation, and the distinction in this case between output and state disappears superficially. However, for consistency of terminology, the distinction is maintained.

The concept of state relates best to systems variable in space and/or time, and to which a control is introduced and an output calculated, but it has wider usage than this.

Example: Network Models Used in Project Planning

For the special network models used in project planning, output and state are the same thing. That is, there are no observability (in the sense of Kalman) issues—the state can be observed directly.

The state relates the "input" control to the output and determines the output uniquely for a given control. Both state and output are *controlled* variables. In this context, only *controlled systems* are implied; that is, the system state and output may be manipulated by careful selection of the control. The term "controlled" implies a causal relation (model) between control, and state and output.

Example

The notions of state, output, and control extend to all levels in a hierarchical system such that, for example, in a building structure, materials, dimensions, geometry, rigidities, ... may all be considered controls, while stresses, strains, curvatures, ... may all be considered states at various levels of the total structure hierarchy.

Exercise 2.30

Consider a materials handling system involving conveyors taking material from a stockpile and depositing it in bins or silos. From the bins, the material is then taken away in trucks. For such a system, what would be suitable choices of state variables and output variables? What variables does the designer manipulate directly in this system?

Exercise 2.31

An experiment on a person. Think of a person as a system. You are to do something to that person (that is, alter the input to the system) and observe the behavior/response of that person (that is, observe the output of the system) (Figure 2.23).

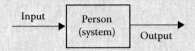

FIGURE 2.23
Person as a system.

The input may be lighting levels, noise levels, pay, reward, training, etc. The output may be productivity, satisfaction, some behavior trait, etc.

Do not do anything that is unlawful or immoral, or something that you would not like done to you. (For example, refrain from violence, bullying, and the like.) Something connected with that person as a worker is preferred.

Examine a range of input values and observe the corresponding output values. Plot the results (that is, input–output values).

There are a number of tricky things that have to be thought through very carefully:

- Make sure that you keep all other potential inputs constant and that you only vary your one selected input.
- Make sure that the system (person) stays the same throughout the experiment. This means more than just the physical thing called a person.
- Be aware that people behave differently if they know that they are being watched or studied (the Hawthorne effect).

Based on your observation of the experiment, suggest suitable models (that is, input–output relationships) for the people–situation that you are examining. Comment on the range of validity of these models.

Comment on general issues that exist in modeling human behavior.

Comment on how long it will be before mankind will have models of human behavior that are of a comparable accuracy to those used for mechanical systems.

2.7.1 Terminology

Generally, the term "control" is used loosely in the literature and by lay persons and practitioners. Particularly at fault is the usage in some computer packages, where package marketing promotes the feature of "control," but really refers to something else, in many cases something no more sophisticated than handling data. In other cases, the term is used to mean placing some limit or containment on something such as costs or keeping this something from growing unacceptably. In general management texts, and lay usage, the term "control" is used in the context of power, dominance, or influence. (Something that is "out of control" is unable to be influenced.) It appears that the word "control," like planning, is used by most people in the sense of the earlier *Alice in Wonderland* quote. The common lay use of terms such as "project control," "project control system," and "cost control system" are misleading; their sloppy usage is inconsistent with the optimum control systems usage preferred in this book.

In this book, the term *control* is used in a strict systems sense, meaning *the actions, input, or decisions selected by the engineer in order to bring about a desired system performance.*

The popular usage of control as stopping something from growing unacceptably or containing something is shown later to be incorporated through the use of constraints in synthesis formulations.

2.7.2 Automatic Control

Automatic control of systems may be done in a number of ways. Of particular importance are the following:

- *Open-loop control*: Information flow is one way. The controls are selected up front, and no follow-up monitoring of system performance and corrective control action is carried out. See Figure 2.24.

- *Closed-loop (feedback) control*: There is a circular transmission of information. See Figure 2.25. Closed-loop control is commonly used in an *error control* form. The control is selected, and feedback occurs based on the difference between actual system performance and desired system performance.

Closed-loop feedback control attempts to nullify the effects of disturbances on system performance and to keep the system performance as desired.

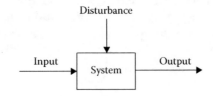

FIGURE 2.24
Open-loop control.

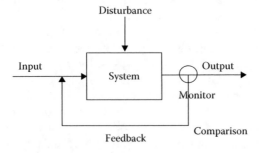

FIGURE 2.25
Closed-loop control representation.

It has the better chance of ensuring that the system's performance goes close to that desired, compared with open-loop control where the project performance may or may not go close to that desired.

In error control, feedback may be in a positive or negative sense. Negative feedback decreases the deviation between actual and desired performance, with the situation of zero deviation being the ultimate target. (Positive feedback increases the deviation.) There is no feedback if there is no deviation.

Both open-loop and closed-loop control include anticipation of the effect of future events/influences (much like the approach adopted in risk management). The control is selected with the view that future system performance is going to be affected by future events/influences.

Feedforward control is a feedback control centered on the input.

Example: Having a Shower

Consider a person having a (bathroom) shower. If the hot and cold taps/faucets are turned on, and the person stands under the water without further adjusting the water flow, this is open-loop control. However, should the person sense the water temperature with his/her hand and adjust the hot or cold up/down until a desirable water temperature and flow is reached, this is closed-loop control. The water temperature and flow are the states or outputs; the control is the adjustment of the hot and cold flows. Clearly, closed-loop control is more likely to give a satisfying shower for the person.

Example: Bullets vs. Missiles

A bullet fired from a gun is an example of open-loop control. Once fired, the final destination of the bullet cannot be influenced. Guided missiles, on the other hand, are examples of closed-loop control. As the missile proceeds toward its target, it senses it current (x, y, z) position. Based on a comparison of its current position with the target coordinates, controls are applied to adjust the direction and velocity of the missile. Missiles usually hit their targets; bullets often miss their targets.

Example: Project Replanning

The corruption that prevents any planned-for project output being obtained exactly can be represented by disturbance. Disturbance leads to the need for continual monitoring and reporting of project progress, and possibly revised controls.

In project replanning, an error is referred to as a variance. It is the difference between planned (or as-planned) and actual (or as-executed, as-built, as-constructed, as-made, …). It is not referring to any probabilistic measure of variability. (In earned value, different definitions of variance are used.)

$$\text{Variance} = \text{Planned} - \text{Actual}$$

The usual convention is for a variance to be negative if, for example, a project is behind production/schedule or, for example, over budget. A variance is positive if the project is, for example, ahead of production/schedule or, for example, under budget.

A project is monitored through its life. Information from monitoring, and a comparison of actual project performance with planned project performance, as well as absolute project performance, is processed into progress reports, which are used as the background to replanning (Figure 2.26).

Figure 2.26 represents a closed-loop control system.

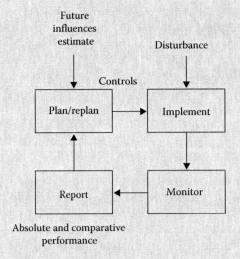

FIGURE 2.26
The replanning process.

Example: Quality

The Deming PDCA (plan, do, check, act) cycle is a version of a closed-loop control system.

Example: Earthmoving Operation

Earthmoving, quarrying, and surface mining operations may be made more efficient by allowing the way they are undertaken to change in reaction to what is happening on the operation. Control in this sense is of the closed-loop or feedback form. Control is carried out in "real time" or "online" typically with instructions issued via radio/telephone communication with the plant operators, or electronically based on GPS (global positioning system) and radio frequency devices.

Controls or inputs to an operation are the actions taken by the operation supervisor to bring about a desired operation output or performance. Example controls might be allocation/redirection of trucks to a different dump site; allocation/redirection of trucks to a different loader; grading/compacting part of a haul road; or using a temporary stockpile.

Example: Gap "Analysis"

In managerial speak, a gap "analysis" establishes how far an organization's practices are from some benchmark, which might, for example, be another organization's practices or some legislated requirements. The gap is the error between actual and desired, and managerial action should then be to do something (implement or change controls) that removes or reduces this gap. Ongoing gap "analyses" represent part of ongoing feedback or closed-loop control. (Note, the term "analysis" is not used here in the same sense as the analysis configuration discussed later.)

2.7.3 Fundamental Configurations

The fundamental systems configurations of analysis, synthesis, and investigation may now be introduced in relation to this terminology.

The *analysis* configuration regards the input (including control) variables to be given, with the only true variables being the state variables. The *synthesis* configuration attempts to assign the control variables so as to give a desired state (direct synthesis) or to optimize some objective (optimal synthesis, optimal control, optimization). Design and planning are particular examples. Both the analysis and synthesis configurations require the *a priori* specification of the system model. However, the *investigation* configuration determines the form of the system model for given input–output characteristics; the terms "black box" or "gray box" may be used in conjunction with investigation, implying a complete or partial lack of knowledge respectively of the organization of the model. The control systems theory embraces these fundamental configurations.

Exercise 2.32

Consider a materials handling system involving conveyors taking material from a stockpile and depositing it in bins or silos. From the bins, the material is then taken away in trucks. What would be a typical

- Analysis configuration?
- Synthesis configuration?
- Investigation configuration?

What information would you needed for the investigation configuration?

3

Some Common System Models

3.1 Introduction

This chapter presents some commonly adopted models. The treatment is not exhaustive, but rather a sample of common models is given.

Outline: The following sections give examples of commonly used models.

3.2 Block Diagrams

One possible representation of systems is that which uses block diagrams. They have visual appeal in understanding how a system functions.

Each block (Figure 3.1) is a graphical representation of cause and effect, or the relationship between input and output.

Typically, such representations contain, in the box, the name of the system, its function, or the mathematical operation that converts the input into an output. An example of the last type is shown in Figure 3.2.

Generally, if the output z is related to the input u through

$$z = hu$$

then h is referred to as a *transfer function* (Figure 3.3).

Most commonly, transfer functions are encountered in the systems literature in conjunction with a classical control systems treatment using Laplace and z-transforms; transforms are taken of the input and output expressions. This is found in electrical, mechanical, and chemical engineering treatments of systems. But the terminology of transfer functions has wider applicability than this.

In structural engineering, the term "transfer matrix" is found. This is a matrix that converts the multiple inputs into the multiple outputs of a structural element. Transfer matrices, as used in structural analysis, have largely been replaced by stiffness matrices (and to a lesser extent flexibility matrices) because

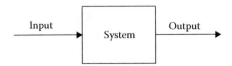

FIGURE 3.1
Block diagram representation of a system.

FIGURE 3.2
Differentiation operation.

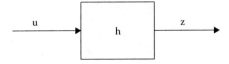

FIGURE 3.3
Transfer function form.

of their wider applicability and better accuracy. Mathematical transformations exist between transfer matrices, stiffness matrices, and flexibility matrices.

3.2.1 Junctions

For systems more complicated than the open loop input–output version of Figure 3.1, it is necessary to introduce junction points, which may mean adding or subtracting information, or branching (Figures 3.4 and 3.5).

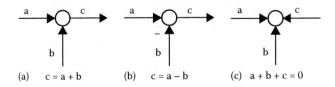

FIGURE 3.4
Junction points example. (a) Addition, (b) subtraction, and (c) general.

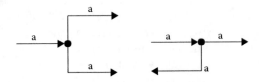

FIGURE 3.5
Branching point examples.

Boxes, for system elements or subsystems that are part of a larger system, together with junction points, permit the representation of subsystem interaction and systems more complicated than the open loop input–output version of Figure 3.1.

3.2.2 Series Subsystems

Consider two subsystems in series (Figure 3.6).
Here,

$$v = h_1 u$$

$$w = h_2 v$$

That is,

$$w = h_1 h_2 u$$

Figure 3.7 is equivalent to Figure 3.6.

3.2.3 Parallel Subsystems

Consider two subsystems in parallel (Figure 3.8).
Here,

$$z_1 = h_1 u$$

$$z_2 = h_2 u$$

$$z = z_1 + z_2$$

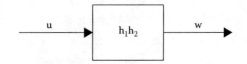

FIGURE 3.6
Subsystems in series.

FIGURE 3.7
Equivalent representation to Figure 3.6.

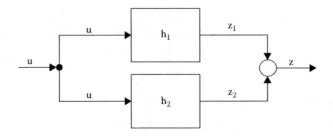

FIGURE 3.8
Subsystems in parallel.

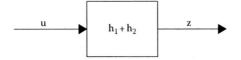

FIGURE 3.9
Equivalent representation to Figure 3.8.

That is,

$$z = (h_1 + h_2)u$$

Figure 3.9 is equivalent to Figure 3.8.

Exercise 3.1

The probability of failure of an element i is given as p_i. Reliability is the probability of no failure. What is the reliability of two elements, 1 and 2, in series? What is the reliability of two elements, 1 and 2, in parallel?

3.2.4 Feedback

Feedback may be indicated by a diagram such as Figure 3.10, where the junction point conventions of Figure 3.4 will apply.

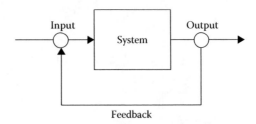

FIGURE 3.10
Feedback.

Exercise 3.2

The book *Catch-22* (Heller, 1964) presents some interesting logic on p. 54, when referring to pilots flying combat missions.

Given the high probability of dying, a pilot would "be crazy to keep flying combat missions." And so in order to be grounded, all the pilot has to do is *ask* [to be grounded]. But on asking to be grounded, the pilot ·cannot be grounded:

> Catch-22. Anyone who wants to get out of combat duty isn't really crazy.

The act of asking implies that the pilot is not crazy.

> There was only one catch and that was Catch-22, which specified that a concern for one's own safety in the face of dangers that were real and immediate was the process of a rational mind. Orr was crazy and could be grounded. All he had to do was ask; and as soon as he did, he would no longer be crazy and would have to fly more missions. Orr would be crazy to fly more missions and sane if he didn't, but if he was sane he had to fly them. If he flew them he was crazy and didn't have to; but if he didn't want to he was sane and had to.

If you were to model the Catch-22 using block diagrams, what might the model look like?

Exercise 3.3

The following is extracted from Heller (1964).

> His [Major Major's father] specialty was alfalfa, and he made a good thing out of not growing any. The government paid him well for every bushel of alfalfa he did not grow. The more alfalfa he did not grow, the more money the government gave him, and he spent every penny he didn't earn on new land to increase the amount of alfalfa he did not produce. Major Major's father worked without rest at not growing alfalfa. On long winter evenings he remained indoors and did not mend harness, and he sprang out of bed at the crack of noon every day just to make certain that the chores would not be done. He invested in land wisely and soon was not growing more alfalfa than any other man in the county. Neighbors sought him out for advice on all subjects, for he had made much money and was therefore wise.

If you were to model this situation using block diagrams, what might the model look like?

Exercise 3.4

The book *Catch-22* (Heller, 1964) presents some interesting logic on pages 110–111, when referring to the Major having visitors.

> The Major didn't want to see anyone in his office while he was there. People who came to see the Major were asked to wait (until the Major left). They were permitted to see him after he had left. But then, the Major would not be there.

If you were to model this situation using block diagrams, what might the model look like?

Exercise 3.5

Supply chain: Consider some work process. Model the supply chain related to this component. Choose something that is doable but not trivial. For example, on a road construction project, you might choose to model the quarried material as it goes from being won, to processed, to transported, to inclusion in the end-product (the road).

Draw a flow (block) diagram of what is happening. Each block will correspond to a distinct activity. Where the work can be done in different ways, there will be more than one block diagram. Include statements of your assumptions.

Do similarly for carbon emissions as a material is processed, formed in manufacture, used, discarded, recycled, etc., along the lines of life cycle assessment (LCA) thinking.

3.3 Black Box

The term "black box" derives from the electrical bakelite boxes usage. In a systems sense, it refers to the input–output relationship without bothering as to what lies inside the system. That is, only input and output variables are considered. The system structure is unknown, of no interest or too complicated.

Treating systems as black boxes enables information to be gained of the system without disturbing or knowing what is happening within the system.

Example: Human Body

Cause–effect: looking at the possible causes, and a patient's symptoms such as temperature, mobility, etc., (effects), some indication can be obtained of a person's state of health. Medicines are then administered (inputs) to give desired outputs (temperature, mobility, etc.), which are assumed to be associated with an appropriate state of health.

Example: Catchments

Black Box

Catchments are sometimes modeled as black boxes. Rainfall (input) and streamflow (output) are measured, and the relationship between rainfall and streamflow obtained is a model of the catchment, and this may be used to predict streamflow for given different rainfalls. Nothing is assumed about the characteristics or behavior of the catchment.

Gray Box

A rainfall–streamflow model may be comprised of component models for infiltration, baseflow, surface runoff, and runoff routing (Figure 3.11). Each of these component models may be based on the physics of the relevant process, but still with parameters to be determined and peculiar for each catchment. Actual site data are used to establish the values of these parameters.

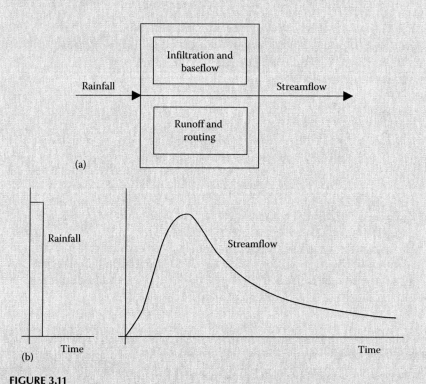

FIGURE 3.11
(a) A general approach for rainfall–streamflow modeling and (b) example catchment hydrograph.

Note that different internal system structures may be possible with the same given external inputs and outputs. Accordingly, there may be a preference for using gray boxes, where some internal structure is assumed.

To develop a black box model, a researcher looks at input–output pairs (alternatively stimulus–response or cause–effect pairs). In some cases, it may be possible to select the inputs. In others, the researcher has to make do with whatever inputs are available.

Example: Catchment, Tension Wire

In developing a model of a catchment, a researcher works with nature's rainfall and runoff. By comparison, in developing a model of a piece of wire or bar in tension, a researcher applies various tensile forces to the wire and measures the associated wire extension (leading to Hooke's law, which is a model of the behavior of the wire in tension).

In addition to observing the input and output values, care has to be exercised that additional extraneous inputs, which can influence output values, do not occur or are allowed for. Some isolation of the system is required when developing a model.

Where multiple inputs and multiple outputs are present, the input–output relationship can be difficult to establish unless inputs are selected one at a time.

Exercise 3.6

For a catchment, what knowledge would you include about the underlying structure of the catchment in a gray box model? What might the component models in Figure 3.11 look like?

How much more reliable than a black box model of a catchment would you expect a gray box model to be in predicting streamflow for any given rainfall?

Example

Pavlov's experiments with animals, treated as black boxes, resulted in the discovery of the conditioned response. Conditioned responses are the result of sequential behavior patterns.

3.4 State Equation Models

State equation or state space models have attracted popularity in representing dynamic systems in modern control systems theory. However, their application is far wider than this. Modern control systems theory adopts real-time, time-domain methods in comparison to earlier control system modeling using Laplace and z-transform methods.

State equation models are advantageous in that they allow common descriptions of all systems (of the class considered), while standard mathematical techniques may be invoked.

The models permit ready understanding of the fundamental systems configurations of analysis, synthesis, and investigation.

The following development is in terms of time, t, as the independent variable. However, the essence of the modeling transfers to the case where a spatial dimension is the independent variable.

Several standard forms of state equation models can be recognized. Systems described by ordinary differential equations and difference equations are referred to as *lumped parameter systems*. The concepts extend readily to systems governed by partial differential equations, referred to as *distributed parameter systems*.

In the following, the state is assumed to take any positive or negative value. However, it is possible to only allow discrete values of the state.

3.4.1 Differential Equation Models

The most common standard state equation form is expressed as a set of first-order equations:

$$\frac{dx}{dt} = f[x, u, t] \tag{3.1a}$$

where

x is an n-dimensional state vector containing components $x_1, x_2, \ldots, x_n$, the state variables

u is an r-dimensional control vector containing components $u_1, u_2, \ldots, u_r$, the control variables

f is an n-dimensional vector function of the arguments shown

t is time

In place of the notation dx/dt, some texts use $\dot{x}$. $x_1, x_2, \ldots, x_n$ define an n-dimensional *state space*, with x_i as coordinates. The state, at any time, may be represented by a point in this space. The state movement over time (the locus of these points) might be called a *state trajectory*.

Note in Equation 3.1a that only states and controls appear on the right-hand side. Derivatives of the state (only) appear on the left-hand side. Parameters or constants are not explicitly stated, even though they may be present.

Associated with Equation 3.1a are initial and/or final conditions, $x(0)$ and $x(T)$, on some or all the state variables. With the investigation configuration, for example, in areas such as parameter estimation and more generally system identification, an additional equation, called the response, output, or observation equation is included:

$$z = h[x, u, t] \tag{3.1b}$$

where
    z is a vector of output or response variables
    h is a vector function of the arguments shown

As well, there may be noise or disturbance in Equations 3.1. Frequently, only the state equations are referred to as the system model, with the output or response equation implied but not stated. It is only for the investigation configuration where the observation equation assumes importance. For the analysis and synthesis configurations, it is generally assumed that the response follows straightforwardly once the state is known.

The linear versions of Equations 3.1a and 3.1b are, respectively (without noise or disturbance),

$$\frac{dx}{dt} = Ax + Bu \tag{3.2a}$$

$$z = Cx + Du \tag{3.2b}$$

where A, B, C, and D are matrices of constants.

The essence of state equations is that they are a series of first-order equations. They can be established from first principles or by reducing a higher-order equation. Any nth-order equation can be reduced to n first-order equations, by introducing n state variables. The state variable form corresponds with the so-termed normal form of the theory of differential equations, and the equations are said to be normalized. Choice is usually available in how the state variables are selected; however, the state equation model is more meaningful if the states are chosen to correspond with something well-known in a physical sense, rather than being some arbitrary mathematics.

Example: Vibrating Lump Mass

The equation for a vibrating lump mass is given by the second-order equation

$$\frac{d}{dt}\left(m\frac{dw}{dt}\right) = F$$

where
  w is the displacement or position, a function of t
  F is the forcing function, a function of t
  m is the mass (a constant)

Let one state variable be displacement or position, that is, $x_1 = w$, one state variable be momentum, that is, $x_2 = m(dw/dt)$, and the control variable be the force, that is, $u = F$. The state might be referred to as the position–momentum pair and is two-dimensional. The two state equations become

$$\frac{dx_1}{dt} = \frac{1}{m}x_2$$

$$\frac{dx_2}{dt} = u$$

These two equations are but a special case of Equations 3.1a or 3.2a. Initial conditions (t = 0) would typically apply on both the state variables. Given the control u, the state is determined at any time t.

For deterministic systems, the state at any time t is the minimal amount of information needed to completely determine the behavior (state) of the system for all other times for any given control. This is sometimes referred to as having a Markov property.

To extend the concept of state to stochastic systems, the state at any time t is regarded as the information that uniquely determines the probability distributions of behavior (state) at all other times. By definition, this describes a *Markov process*. This definition for state in stochastic systems is a basic assumption and implies a form of dependence between adjacent states. In general, the system may not have states with these properties, but the assumption allows analogous treatments between the deterministic and stochastic cases. The assumption enables a way forward that would otherwise not be possible if full stochastic dependence of states was employed.

Example: Bernoulli–Euler Beam

The constitutive relationship for a beam is a fourth-order differential equation,

$$\frac{d^2}{dy^2}\left[D(y)\frac{d^2w(y)}{dy^2}\right]=q(y)\quad y^L\le y\le y^R$$

where the notation is given in Figure 3.12. Consider from a designer's viewpoint: for a given loading q (constant), the designer is able to choose the flexural rigidity, D, of the beam. That is, the control u = D.

This fourth-order equation may be reduced to four first-order equations through selection of the following four states:

x_1 (the deflection w)
x_2 (the rate of deflection or slope, dw/dy)
x_3 (the internal moment—proportional to the second derivative of w)
x_4 (the internal shearing force—proportional to the third derivative of w)

That is,

$$x_1=w\quad x_2=\frac{dw}{dy}\quad x_3=D\frac{d^2w}{dy^2}\quad x_4=\frac{d}{dy}\left(D\frac{d^2w}{dy^2}\right)$$

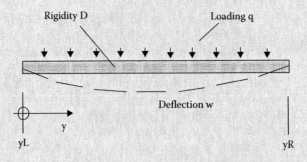

FIGURE 3.12
Beam notation.

These, not coincidentally, correspond with the static and kinematic boundary conditions on the beam at its ends. The state equations become,

$$\frac{dx_1}{dy} = \frac{dw}{dy} = x_2$$

$$\frac{dx_2}{dy} = \frac{d^2w}{dy^2} = \frac{x_3}{u}$$

$$\frac{dx_3}{dy} = \frac{d}{dy}\left(D\frac{d^2w}{dy^2}\right) = x_4$$

$$\frac{dx_4}{dy} = \frac{d^2}{dy^2}\left(D\frac{d^2w}{dy^2}\right) = q$$

and are a special case of Equations 3.1a and 3.2a, with y (spatial dimension) replacing t (time or temporal dimension).

For a given control u, loading q (constant), and boundary conditions, the state is completely determined at any y. The state (at any y) contains the minimal amount of information required to determine the state at some other position y', for any given control.

The state represents the internal behavior of the beam. The outward behavior follows straightforwardly from knowledge of the state and control. The state concept exploits the basic composition of the structural system equations.

The output or observable variable, z, equals w or x_1 directly.

The three basic relationships (equilibrium, compatibility, and constitution) in the earlier beam example are readily identified: the last two equations comprise the equilibrium relationships, and the first two comprise compatibility and constitution. The states x_1 and x_2 are generalized displacements while the states x_3 and x_4 are generalized forces.

3.4.2 Difference Equation Models

A state equation form equivalent to Equation 3.1a exists for the case where the independent variable t takes on only discrete values. The corresponding system is referred to as a *discrete (data) system* as opposed to a *continuous (data) system*, where the variables are functions of the continuous independent variable t. (Where the variables are in the form of a pulse train or a numerical code, the system is often referred to as a *sampled data system*. A *digital*

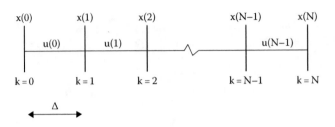

FIGURE 3.13
Discretization notation.

system implies the use of a digital computer or a digital sensing element in the system. Both sampled data and digital systems are discrete data systems.)

The standard first-order state vector difference equation form, in analogy with Equation 3.1a, and output equation, in analogy with Equation 3.1b, are

$$x(k+1) = F[x(k), u(k), k] \quad k = 0, 1, ..., N-1 \tag{3.3a}$$

$$z(k) = h[x(k), u(k), k] \quad k = 0, 1, ..., N \tag{3.3b}$$

Equation 3.3a may be interpreted as a sequence of transitions from the state at k to the state at k + 1, k = 0, 1, ..., N − 1. With information only available on the states at discrete points, the control u(k) is considered to be maintained constant during each interval and changed in a step manner at these points (Figure 3.13).

The state is assumed to have Markov properties; that is, the state at k + 1 depends only on the immediately previous state x(k) and control u(k).

Example: Vibrating Shaft

Consider an elastic vibrating shaft, formed by the assemblage of N concentric rings (Figure 3.14).

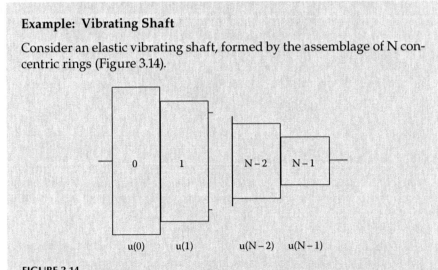

FIGURE 3.14
Shaft.

For segments of constant diameter, the states (rotation ϕ and $JG(d\phi/dy)$) are related from one side of a segment to another through

$$\begin{bmatrix} x_1 \\ x_2 \end{bmatrix}_{k+1} = \begin{bmatrix} 1 & L/JG \\ r^2\gamma\omega & 1 \end{bmatrix}_k \begin{bmatrix} x_1 \\ x_2 \end{bmatrix}_k$$

where
  L is the length of the segment
  JG is the torsional rigidity
  r is the radius of gyration
  γ is the mass per unit length
  ω is the circular frequency of vibration

From a design viewpoint, the controls relate to the geometrical properties JG and r.

A discrete representation of a system in the form of Equations 3.3a and 3.3b may be suitable not only for the case where the system is inherently discrete but also as an approximation to a continuous system, as an aid to computations. This latter case is particularly relevant for the stochastic control case where the discrete version is conceptually simpler while avoiding the heavy rigor required in the continuous stochastic case. If time (or spatial dimension) is discretized while keeping the probability space continuous, the stochastic processes $x(t)$ and $u(t)$ now become random sequences and are completely defined by their finite-dimensional distributions.

Using the finite-difference relation at k,

$$\frac{dx}{dt} = \frac{x(k+1) - x(k)}{\Delta}$$

over [0,T] partitioned into N intervals of size $\Delta = T/N$, Equation 3.1a becomes the difference equation (Equation 3.3a). The behavior of the discrete and the behavior of the original continuous models are assumed to be similar as the interval Δ goes to zero. Boundary conditions on Equation 3.3a are expressed at $k = 0$ and $k = N$. The interval Δ need not be assumed constant.

Example: Bernoulli–Euler Beam

Consider the Bernoulli–Euler beam state equations developed earlier. The beam length [0,L] is divided into N equal intervals $\Delta = L/N$ such that at any position $y = k$, $k = 0, 1, ..., N$, the continuous form of the state equations may be restated in the discrete form,

$$\begin{bmatrix} x_1(k+1) \\ x_2(k+1) \\ x_3(k+1) \\ x_4(k+1) \end{bmatrix} = \begin{bmatrix} x_1(k) \\ x_2(k) \\ x_3(k) \\ x_4(k) \end{bmatrix} + \Delta \begin{bmatrix} x_2(k) \\ x_3(k)/u(k) \\ x_4(k) \\ q(k) \end{bmatrix}$$

Boundary conditions apply at $k = 0$ and N.

Structural engineers will be familiar with the transfer matrix, which can be obtained from these equations:

$$\begin{bmatrix} x_1 \\ x_2 \\ x_3 \\ x_4 \end{bmatrix}_{k+1} = \begin{bmatrix} 1 & \Delta & & \\ & 1 & \Delta/u & \\ & & 1 & \Delta \\ & & & 1 \end{bmatrix} \begin{bmatrix} x_1 \\ x_2 \\ x_3 \\ x_4 \end{bmatrix}_{k}$$

And from this, you can obtain the stiffness and flexibility matrices for the beam. Transfer matrices have fallen from favor, but stiffness matrices have not.

Linear versions, equivalent to Equations 3.2, can also be given:

$$x(k+1) = A'x(k) + B'u(k) \tag{3.4a}$$

$$z(k) = C'x(k) + D'u(k) \tag{3.4b}$$

where A′, B′, C′, and D′ are matrices.

3.4.3 Partial Differential Equation Models

The earlier concepts extend readily to systems governed by partial differential equations, where the independent variables are time and space.

Three standard partial differential state equation forms may be recognized, two of which reduce to the standard lumped parameter form on simplification (Carmichael, 1981).

For manipulation purposes, discretization of all or some of the independent variables of time and space might be carried out.

3.4.4 Algebraic (Nondifference) Equation Models

By far, the most common models, and the most preferred by people, are the simple algebraic versions. Generally, algebraic models take the form

$$F(x, u) = 0 \tag{3.5}$$

where
u is an r-dimensional control vector
x is an n-dimensional state or response vector
F is a vector-valued function of the arguments shown (note: a different F is implied from that in Equation 3.3a)

The distinction between control variables and behavior variables might be dropped in synthesis studies.

Algebraic models correspond in mathematical terms to the so-called *static models* of control theory. This is compared with the earlier models, which contain time and are *dynamic models.*

Both the algebraic and state space forms may be converted to the "input–output" forms of classical control theory.

Example: Structure

Consider a matrix stiffness representation of a structure:

$$R = K\Delta$$

where
R is the structure nodal loading vector
Δ is the structure nodal displacement vector
K is the structure stiffness matrix with entries such as EI/L, EA/L, ...
where E is the modulus, I moment of inertia, A area, and L length

A distinction between behavior-type variables and controls may be drawn. Controls relate to materials and geometry, which are at the disposal of the designer to vary, and include the elastic modulus, and first and second moments of area. Behavioral variables relate, for example, to the displacement.

Exercise 3.7

What value do you see in having representations (such as Equations 3.1 through 3.5), which are standard systems models? Or would it be better for every system to have its own unique representation?

Is representing systems by a set of first-order differential equations or difference equations restrictive in its range of applicability?

3.5 Other Forms

Other forms of common models that many people group within systems studies are deliberately omitted here, because other books are readily available covering them.

3.5.1 Networks

In simple terms, a network is a collection of *links* joining *nodes*. Such representations are popular for modeling traffic and transport routes, water and electrical distribution, framed structures, logical ordering, and the like. A special form of *directed network* is used in project planning.

3.5.2 Queuing Models

Queuing or waiting line models are applicable wherever there is a customer–server relationship, where the terms "customer" and "server" can apply to nearly anything including people, equipment, and materials. The associated theory is readily applied to many engineering operations.

3.5.3 Trees

Trees are a form of representation that mimic natural trees. They are popular because of their pictorial form. Consider organizational structures, work breakdown structures, and similar, and how they help understanding. Two common examples used in reliability engineering are fault trees and event trees, while decision trees are used in decision making.

3.5.4 Simulation

Simulation may involve model development as well as analysis, or if the former has already been done, analysis alone.

In many systems studies, and particularly where complicated models are involved, it is not possible to develop closed-form expressions connecting input and output. As well, there may be no developed numerical calculation procedure. Simulation, in such cases, may be the only viable form of analysis; approximations in order to make other approaches tractable may not be acceptable.

Simulation might be classified as to whether it applies to static/dynamic, stochastic/deterministic, or discrete/continuous models. There are numerous versions of simulation, including

- Monte Carlo simulation
- Discrete event-oriented simulation
- System dynamics
- Trainers, games

4

Fundamental Configurations Relating to Systems

4.1 Introduction

Configurations relating to systems may be classified into three types:

- Analysis
- Synthesis
- Investigation

Consider a system represented as in Figure 4.1a. Let the control input be A, the model of the system B, and the output C, as in Figure 4.1b. For other inputs, known or estimated, the fundamental configurations become

Analysis	Given A and B, obtain C
Synthesis	Given B and C, obtain A
Investigation	Given A and C, obtain B

In each, something different is known, and it remains to obtain the unknown.

Synthesis and investigation are known as *inverse configurations*. In general, their results are *nonunique*, whereas the result of the analysis configuration is generally unique. Nonuniqueness in the synthesis and investigation configurations might be removed through additionally including some optimality measure.

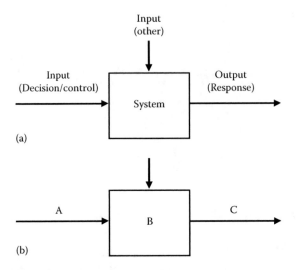

FIGURE 4.1
System representation. (a) Input and output descriptions, (b) input and output as symbols.

Exercise 4.1

A structural designer chooses spans and sizes, and to a lesser extent material properties, for structural members in order that s/he ends up with a structure that behaves in the correct way. Spans, sizes, and material properties are control/decision/design variables, while the behavior is the output or response. That is, the structural designer is working with a synthesis configuration. Why then is analysis the main approach used by such designers?

Exercise 4.2

What configuration type is planning, that is establishing how work will be done, in what order, and with what resources?
  In planning a project, what are the inputs (decisions/controls) and what are the outputs?

Exercise 4.3

Of the three configuration types, which involves the most creativity? Why?

Outline: The following sections give an overview of the fundamental configurations relating to systems. The concepts of controllability and observability are interesting, though usable results in the literature are lacking. The concepts are given as food for thought.

4.2 Analysis

Approaches that fit within analysis include

- Simulation (but can also combine with modeling)
- Prediction, forecasting (after a model is established)
- Stability
- Sensitivity, "what if"
- Economic appraisal (feasibility, and preference)
- Critical path method (CPM)
- Reliability
- Queuing theory
- Fault trees, event trees
- Numerous specialty methods

The end result of analysis is knowing the behavior (typically external behavior or output) of the system. Analysis configurations, in principle, generally lead to a *unique* value for the output.

Analysis is deliberately omitted from this book, because analysis comprises the majority of the engineering literature, and many books are available on every possible form of analysis.

4.2.1 Simulation

Simulation may be described as a form of numerical experimentation, but is analysis in that numerical inputs are used to produce numerical output. In some versions, simulation may also contain some modeling, in addition to analysis.

There are many forms of simulation, including Monte Carlo simulation, system dynamics, and discrete event-oriented simulation.

4.2.2 Prediction, Forecasting

Typically based on historical data, a forecasting or prediction model is developed. However, once developed, the model is used in an analysis sense.

Writers may blur the distinction between modeling and analysis, and include both together in one treatment.

4.2.3 Stability

Stability is a characteristic of a system where, for example, the response to a stimulus (input) reduces when the stimulus is removed. Instability implies that the response does not reduce when the stimulus is removed—the response may stay the same or increase. Some might also argue for the term "unstable" to include the case where the response does not increase.

Here, the stimulus is an input to the system from which the response is observed. As such, an examination of a system's stability is closely linked to the analysis configuration. To ascertain whether a system is stable, a stimulus is applied/removed to the system model and the response calculated. Depending on the nature of the response, the system is considered stable or unstable.

4.2.4 Sensitivity

Sensitivity may be viewed as the change in the system response to changes in either

- The input or
- The system makeup including system parameters, boundary conditions, initial conditions, and so on

As such it fits an analysis configuration.

Commonly, it might be called a *what if* analysis. For example, *What if the input is changed or what if a system parameter is changed?*

Strictly, sensitivity analysis only deals with *small changes* to the input or system makeup (whereas "what if" can be any scenario). For such changes, the response is examined and compared with the original response. This comparison establishes whether the system is sensitive (to input or system makeup changes). Response changes of magnitude bigger than the perturbing changes or of an order of magnitude bigger than the existing response would commonly imply high sensitivity. Small magnitude changes in the response would imply low sensitivity.

Refer to Appendix A.

4.2.5 Economic Appraisal

In economic appraisal (feasibility and preference), future benefits and future costs (input) are commonly discounted to present-day values (output in present worth or net present value approach) or annual values (output in annual worth approach), or manipulated to give an interest/ discount rate (output in internal rate of return approach), or manipulated to see how long before the benefits exceed the costs (output in payback period approach).

For feasibility of an investment, these outputs are compared with benchmarks. For preference between investments, these outputs are compared among the alternatives.

4.2.6 Critical Path Method (CPM)

Planners use a network model to represent the logical order in which activities are to be done on projects. Planning strictly belongs to the synthesis configuration, but almost universally planners use analysis, coupled with iterations, because it is simpler to work this way. The inputs are activity durations and resources (people, equipment—which can be given a cost value), while the outputs are activity start and finish times, project duration, and resource usage (which can be given a cost value).

4.2.7 Reliability

Reliability, in a general sense, refers to a system's (something's, someone's) ability to perform a specified task for a specified time in a specified environment (that which is outside the system). It is measured as a probability. A system's calculated output is compared with its capacity and the reliability, or probability of no failure, calculated.

4.2.8 Queuing Theory

Queuing or waiting line theory deals with various customer–server relationships, where customers and servers are general terms covering people, equipment, workplaces, etc. Typically, for given customer and server information, a queuing analysis gives as output waiting times, server utilizations, queue lengths, and similar.

4.2.9 Fault Trees, Event Trees

Fault trees address the question: How can failure occur? What could happen that would give rise to or cause this failure? Fault tree analysis starts from a (root) fault and progresses to what might cause the fault, and then to what might bring about these causes and so on. Event trees address: What could happen or follow if a given event occurs? Event tree analysis traces subsequent events leading from an initiating event.

4.2.10 Other Forms of Analysis

There are uncountable methods of analysis available, ranging from quite sophisticated mathematical forms (involving matrices, differential equations, etc.) to crude "gut feel" or "seat-of-the-pants" qualitative varieties.

The end result of analysis may be quantitative or qualitative. The quantitative forms may give answers on a ratio scale. Qualitative forms may be restricted, for example, to ordinal scales (for example, high, medium, low).

Analysis occupies most of the technical literature, because it is simpler to deal with than synthesis or investigation. To illustrate this, when working with synthesis or investigation, it is frequently easier to convert synthesis and investigation configurations into an analysis configuration, which is adjusted iteratively, rather than work with synthesis or investigation directly.

4.2.11 Garbage In–Garbage Out

Garbage in–garbage out (GIGO) is a reference to the output being silly if the input is silly.

A related trap is one that beginners fall into, namely, quoting output to a higher level of accuracy than the input level of accuracy. Computers are able to provide answers to many decimal points, but the underlying accuracy is no better than the accuracy of the inputs.

4.2.12 Modeling and Analysis

Frequently, the technical literature is loose in its distinction between modeling and analysis. For example, simulation can involve both the development of a model, often specific to the situation, and analysis; queuing or waiting line theory similarly involves the development of models and analysis; and prediction or forecasting may firstly involve model development. However, as presented in the technical literature, it appears seamless between the modeling and the analysis. This is exacerbated by the practice, for example, of adjusting a model until the output of the analysis fits some preconceived view of what the output should look like.

It is argued that a better discipline of thought, and one that can assist in situations previously not encountered, is to focus on modeling and the three fundamental configurations as separate entities.

Exercise 4.4

Blood samples, tissue samples, etc., extracted from patients are sent to pathology laboratories for testing. This testing is commonly called analysis. But is it analysis in the sense defined earlier?

Exercise 4.5

In undergraduate engineering programs, the majority of curriculum time seems to be spent on teaching analysis tools. Some time is devoted to developing models from first principles and empirically. Very little time is devoted to synthesis (and investigation). Why is this so, considering engineers, when they graduate, will largely be working with synthesis configurations?

4.3 Synthesis

Approaches that fit within synthesis include

- Design: optimal design
- Optimization
- Optimal control theory
- Decision theory: decision making
- Planning
- Management
- Risk management
- Value management
- Work study

The synthesis configuration is essentially a converse to the analysis configuration. Given a certain (desired) behavior and model, what are the controls/decisions that produce this behavior? Generally, a certain behavior is realizable with many controls/decisions. That is, the result is *nonunique*. Further requirements such as extremization (maximization or minimization, implying optimization) of an *objective*, for example, minimum weight or minimum cost, are needed to make the result unique.

NOTE: *The term "objective" is perhaps the most popular to denote that entity by which the best solution is chosen. However, different disciplines may use different terms such as "optimality criterion," "performance index," "payoff function," "figure of merit," "merit function," "goal," "cost function," "design index," "target function," "criterion and mission."*

In this book, a rigorous definition and usage of the term "objective" is attempted. Note that the term "objective" is used very loosely by most people. For example, a project's objectives are commonly said to be the end-product of the project, or alternatively a project's objectives and its scope are talked of interchangeably. When reading the term "objective" in non-system's work, be aware that it has a very imprecise and loose meaning, and possibly different to that given here. Banish from your mind the lay or dictionary meaning of the word.

Dealing with a synthesis configuration could be expected to be more complicated and more difficult than dealing with an analysis configuration. In some cases, the end result of analysis can even be arrived at intuitively, whereas obtaining an end result in synthesis involves more depth of understanding.

Freedom is the absence of constraint.

Dictionary Meaning

Freedom is the absence of choice.

Sufi Aphorism

4.3.1 Confusion in Terminology Usage

Generally, there is confused interchangeable usage of the terms "design" and "planning." People get away with sloppy terminology usage because design and planning belong to the same synthesis configuration and hence use the same mathematical methods. Typical wrong usages include people calling site layout selection on construction sites as planning, and landscapers' selection of plants and garden borders as landscape planning. This is not helped by some people's preference for calling technical drawings by the word "plan"; such drawings are the outcome of the design process. Project personnel talk of a "project management plan" (PMP), which includes administrative procedures, contract issues, risks and opportunities, and various other miscellaneous matters unrelated to planning.

It is also remarked that many designers and planners do not understand the synthetic nature of their job. These designers and planners are unaware of and do not understand the components of the synthesis configuration. Design to some people is satisfying a code of practice, and so the "distinction," for example, between a university structural analysis course and a structural design course is that the latter mentions a code of practice.

Much terminology usage is in the Alice in Wonderland sense.

Exercise 4.6

Many industry and lay people use the term "planning" very loosely and to mean nearly anything they want it to mean. For example, a person may use interchangeably the terms "design" and "planning" without knowing the difference. Focus on what planning really is and use the term in its correct sense. This will require discipline on your part and to translate what other people are saying when they use the term "planning" incorrectly. Many activities that people include under the banner of "planning" are not planning activities. This is particularly so in managerial parlance.

The term "plans" to denote drawings is well established, but this is not too much of a concern. It is all the other uses that people adopt that shows confusion as to what planning is. And then, there are uses such as town planning, which is really town design. Is a project development plan (PDP) a plan? Count the number of uses of the term "planning" you come across in your organization. It is suspected that few actually relate to real planning. The term "planning" sounds nice and impresses and so is used. The distinction between "means to end-product" and the "end-product" is crucial to understanding the difference between planning and design. One concerns how you get there (planning), and the other concerns what it is when you get there (design). So, for example, if a building is the end-product, design relates to choice of shape, color, size, etc.; planning relates to the activities and resources used in order to bring about that building. The two are connected, but are different.

Reflect on the difference between design and planning, and what planning actually is.

Consider "town planning" and "landscape planning." Are these planning activities or design? What about a "safety plan," "quality plan," "risk plan," "management plan," "business plan," etc.? If you look hard enough, you will see the word "plan" or "planning" everywhere, often being used because it sounds good, but not used in the correct sense. Unfortunately, dictionaries give lay meanings, which are usually useless for technical words. So avoid dictionary and lay usage for technical terms. This means that you have to be bilingual.

4.3.2 Conversion to an Iterative Analysis Form

Because of the degree of difficulty of working with the synthesis configuration compared to the analysis configuration, much synthesis is translated to iterative-analysis.

The control/decision/design variables are evaluated based on an objective. And the behavior is obtained for each set of control/decision/design values. The engineer hopes that by adjusting the initial guesses for the control/decision/design variables, the behavior resulting from the analysis will become more favorable and also the value of the objective will improve (decrease or increase, as appropriate). However, there is no guarantee of this, although the engineer's expertise and knowledge will usually head the iterations in the preferred direction. Where the engineer has worked with such systems before, the engineer's first guess of values for control/decision/design variables may be close to "optimum," and no iterations may be required; nevertheless, the synthesis is still being approached via analysis.

Exercise 4.7

When an architect develops a drawing for a building, is that architect engaged in a synthesis-style process or some other process?

What are the control/decisions able to be exerted by the architect? What are the outputs or responses of interest to the architect?

4.3.3 On-Line/Off-Line

Synthesis can work in two modes:

- Off-line
- On-line or real time

With off-line, synthesis is done once in isolation. With on-line, the synthesis process is continuously updating the controls/decisions as new data come to hand.

4.4 Investigation

Approaches that fit within investigation include

- Parameter estimation
- State estimation
- System identification: characterization
- Black box

All except state estimation belong to modeling. They are attempting to establish input–output relationships for systems.

Modeling is relevant to all disciplines, not just engineering and the physical sciences. The social sciences, economics, medicine, etc., all have an interest in developing quantitative models for their systems.

System identification ranges from the complete black box form where nothing is known about the system model, to the gray box where typically only parameters in the model remain to be estimated (parameter estimation).

Strictly, the version of the model obtained through using input–output pairs is only valid for that range of pairs, although people, as part of human nature, tend to want to generalize the validity of the model beyond the range for which it is calibrated or derived.

The black box form, and in particular the determination of the structure of a model, is very difficult to work with. By far, the gray box and parameter estimation forms are easier and tend to give more realistic models. What lies inside the black box always remains unknown.

4.4.1 State Estimation

State estimation refers to filtering a system's output in order that some best estimate of a system's state is obtained. The filtering typically removes noise from the output measurements or observations. State estimation assumes a known system model.

Mathematically, for dynamic systems, parameter estimation and state estimation might be treated similarly through the device of calling a parameter a state and developing an equation for the change in the parameter's values over time.

State estimation may be used as a precursor to predicting or forecasting future behavior.

4.4.2 Natural/Artificial Input

Investigation may be approached through the use of natural system input, or the system may be artificially stimulated under deliberate experiments. Typical artificial stimuli include (im)pulses, step functions, ramp functions, and sinusoidal functions.

During investigation, the system ideally should be isolated from inputs/ stimuli other than those intended by the observer. In deliberate experiments, this may be possible, while it may be not so in many natural, field, or social systems.

4.4.3 Roots

Many of the techniques used in investigation have common roots in optimization. Something is being maximized or minimized, as in optimization, in order that the behavior of the model best follows that observed in the system. Hence, what some people refer to as optimization is more correctly called investigation.

4.4.4 On-Line/Off-Line

Investigation can work in two modes:

- Off-line
- On-line or real time

With off-line, system data are firstly collected and then the investigation is carried out. With on-line, the investigation process involves continual updating (of, for example, the system model) as new data come to hand. Typically, with on-line investigation, the investigation process follows a prediction then correction format.

Exercise 4.8

When you visit a doctor, s/he examines your symptoms (output, behavior, response). There may be some questions asked about the course of events/activities leading to the illness or injury (input).

From a catalogue of historical and experimental models learned by the doctor at medical school or in technical journals, the doctor attempts to establish the link between input and output. Does this fit an analysis, synthesis, or investigation configuration?

The doctor may then prescribe some medicine (input) in order to alter the symptoms (output). Does this fit an analysis, synthesis, or investigation configuration?

Exercise 4.9

What elements of the practices of managers would you classify as belonging to analysis, synthesis, or investigation configurations?

Exercise 4.10

In order to motivate someone in the workplace, it is argued that you firstly have to establish their needs and then devise incentives that satisfy those needs.

What configuration type does the establishment of a person's needs belong to?

What configuration type does predicting what the behavioral effects of an incentive belong to?

Exercise 4.11

An experiment (Hooke's law) carried out in high school physics is to suspend a mass from the end of wire (fixed to the ceiling at the other end). As extra mass is added, the extension of the wire is recorded.

This gives a series of mass–extension pairs, which, when plotted, follow approximately a linear relationship.

Why would this belong to an investigation configuration? What is the model or input–output relationship being found?

4.5 Controllability and Observability

The concepts of controllability and observability appear to be first developed in a mathematical sense by Kalman in the 1960s.

4.5.1 Controllability

Without exploring the mathematics, a system is said to be uncontrollable if not every state variable is affected by the control inputs. That is, controllability requires that every state of the system be affected by the input/control. In a controllable system, it is possible to move to a desired state through appropriate choice of inputs/controls.

4.5.2 Observability

Observability requires that every state of the system affect the measured output. A system is observable if all its states are derivable from the system's outputs. If states change without affecting the output, the system is unobservable. An unobservable system cannot be identified.

Exercise 4.12

Is the human body unobservable? That is, by observing the outward behavior of a person, can the inner state be established? (Note, this exercise is not looking for causal input–output relationships, but rather the relationship between external output and internal states.) Consider with regard to

- Motivation
- Physical illnesses (nosology is the branch of medicine concerned with the classification of diseases)
- Mental illnesses

The exercise is looking at the connection between outputs and states of systems, and in particular for the very complicated system called a human.

Have you tried self-medical diagnosis? There are some books available and programs on the web. You describe your symptoms, and the program diagnoses your illness. The programs are typically based on expert system shells. Doctors do similarly, but rather based on their knowledge of nosology and experience. How accurate are such diagnoses? Diagnosis is attempting to guess the (internal system) state from external outputs (and possibly knowledge of some inputs).

Do you watch TV shows on marital and family issues? People tell the host their marital and family issues, and s/he diagnoses the cause. (Of course, there could be many causes, but as long as the host talks positively and with (pretend) authority, people believe her/him, even though s/he possibly gets the diagnosis wrong.) The host is attempting to guess the (internal system) state of a marriage from external outputs (and possibly knowledge of some inputs).

Exercise 4.13

Is a country's economy uncontrollable (in spite of what economists and politicians say)? There are a number of economic measures used to characterize economies or an economy's behavior such as GDP (gross domestic product), unemployment, deficit, business failures, business starts, loans, inflation, ... A government can change the economy by changing interest rates, altering gold reserves, money supply, legislation, taxes, etc. There are many inputs to an economy, and there are many outputs (indicators of behavior) to an economy, but it seems (to a noneconomist) that the connection between the two is very ill-defined. This is most notable when you hear economists and politicians disagreeing on what are the best inputs to change in order to get more desirable outputs. That is, the model of a national economy is perceived as being very crude. What are your views?

List the variables that countries use to describe their economies (for example, inflation, employment, ...). These are the outputs.

List the variables that countries can adjust in order to change their economies (for example, interest rates, currency value, ...). These are the inputs.

Now ask: what is the link between the inputs and outputs? Can you adjust the inputs to get a desired set of outputs—a qualitative answer is all that is required.

Extra-country inputs that can affect a country's economy include oil prices, exchange rates (import and export costs), globalization, instability of world peace, epidemics, foreign investment, economies of trading partners, global supply and demand, ... Intra-country inputs include interest rates, altering gold reserves, wages, taxes, political stability, unemployment, government spending, savings, ... Which is stronger—the extra-country inputs or intra-country inputs?

5

The Synthesis Configuration

5.1 Introduction

This chapter looks at a range of popular approaches that fit within the synthesis configuration, including design, optimal design, optimization, optimal control theory, decision theory, decision making, planning, and management. Numerous specialty approaches may be found connected with examples from other chapters.

Some of the following discussion, which has an emphasis on design, has equal applicability to planning, management, and other approaches within the synthesis configuration.

The synthesis configuration is essentially a converse to the analysis configuration. Given a certain (desired) behavior and model, what are the controls that produce this behavior? The end result of synthesis is knowing the system controls, or actions, or decisions that need to be taken.

Synthesis could be expected to be more complicated and more difficult than analysis. In some cases, an analysis result can be obtained intuitively, whereas this is only the case in elementary synthesis when working with simplified models.

5.1.1 Objective

Synthesis configurations, in principle, lead to *nonunique* results; a certain behavior is realizable with many controls. An additional measure (objective) is required in order to be able to distinguish one result from another. Extremization (maximization or minimization, implying optimization) of this measure is needed in order to establish a unique (optimum or best) result.

By the incorporation of an objective, it could be expected that in most cases, the synthesis will lead to one preferred (unique, best, or optimum) control. And generally, it is assumed that the engineer is looking for the optimum; *hence the terms "synthesis" and "optimum/optimal synthesis" tend to be used interchangeably. Similarly for design and optimum/optimal design. Where the optimum nature of something is wished to be emphasized, then the terms optimum or optimal might be appended.*

NOTE: The term "objective" is perhaps the most popular to denote that entity by which the best solution is chosen. However, different disciplines may use different terms such as *optimality criterion, performance index, performance measure, payoff function, figure of merit, merit function, goal, cost function, design index, target function, criterion, mission,* and *aim* (Carmichael, 1981). Where the objective is expressed mathematically, it may be referred to as an objective function.

In this book, a rigorous definition and usage of the term "objective" is attempted. Note that the term "objective" is used very loosely by most people and extremely loosely in the managerial literature; for example, a project's objectives are commonly said to be the end-product of the project, or alternatively a project's objectives and its scope are talked of interchangeably. When reading the term "objective" in nonsystems' work, be aware that it has a very imprecise and loose meaning, and almost certainly different to that given here. The term "objective" is used in the Alice in Wonderland sense by most people.

Banish from your mind lay and dictionary meanings of the term and those given in the popular management literature, including project management.

5.1.2 Constraints

Within the synthesis configuration, there also exist constraints, which restrict the number of possible outcomes. A control/action/decision that satisfies the constraints is referred to as admissible. Only from among those admissible controls is the best control chosen. Sometimes constraints help by restricting the number of possible controls that need consideration, sometimes constraints hinder by making the calculations more difficult.

5.1.3 Variables

The variables manipulated by the engineer in synthesis are referred to as control variables. They are inputs chosen at the discretion of the engineer (as distinct from other system inputs that are not able to be influenced by the engineer). Synonymously, the terms "decision" or "design variables," or "action" might be used.

Outline: The chapter shows how people approach synthesis via the much easier analysis configuration. Synthesis can be given a neat framework if couched in optimal control terms. The treatment covers design, planning, management, risk management, work study, value management, and constructability as synthesis configurations.

Risk management, work study (including the subsumed reengineering), value management (equivalently value engineering, or value "analysis"), and constructability are particular examples of (subsumed by) synthesis via iterative analysis, with each using its own terminology or jargon. If an understanding is obtained of synthesis via iterative analysis, then an understanding of risk management, value management, constructability, and work study will follow naturally.

Synthesis Examples

The composing of *music* belongs to the synthesis configuration. The outcome (the final composition) is nonunique, as evidenced by the number of tunes/songs that exist. The choice of the final composition presumably is determined by selecting that which maximizes some feelings within the composer and is constrained by music genre, existing tunes/songs (in order to avoid copyright, and to appear creative), and the skills of the composer.

Art similarly is synthesis. The outcome (the final painting, sculpture, ...) is nonunique, as evidenced by the number of paintings, sculptures, ... that exist. The choice of the final form of the art presumably is determined by selecting that which maximizes some feelings within the artist and is constrained by the art medium, existing artwork (in order to avoid copyright, and to appear creative), and the skills of the artist.

5.1.4 Decision Support

The term "decision support" is a generalist term, used to describe anything that is connected with decision making, including optimization, data bases, and pictorial representations.

5.2 Conversion to Iterative Analysis

Because of the degree of difficulty of synthesis compared to analysis, much synthesis is treated in an iterative-analysis fashion much like Figure 5.1.

The control/decision/design variables are evaluated according to the objectives. And the behavior is established for each set of control/decision/design values. The engineer hopes that by adjusting the initial guesses for the control/decision/design variables, the behavior resulting

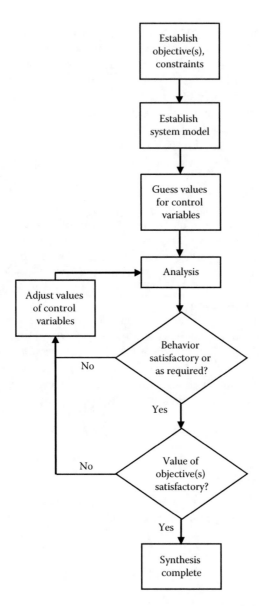

FIGURE 5.1
Iterative-analysis version of synthesis (definitional step not shown).

from the analysis will become more favorable and also the value of the objectives will improve (decrease or increase as appropriate). However, there is no guarantee of this, although the engineer's expertise and knowledge will usually head the iterations in the preferred direction. Where the engineer has worked with such systems before, the engineer's first guess of values for control/decision/design variables may be close to "optimum" and no iterations may be required; nevertheless, the synthesis is still being approached via analysis.

5.2.1 Steps

More broadly, the steps in such an approach as Figure 5.1 might be described as

- Definition
- Objectives and constraints statement
- Alternatives generation
- Analysis and evaluation
- Selection

These steps are fleshed out as follows.

Feedback and iterative modification, for clarification and refinement purposes, may occur within any of the steps, and within and between levels of any system hierarchy. Synthesis becomes an iterative trial-and-error process. This iteration is not inherent in synthesis, but occurs because of the analysis-based mode of attack.

Exercise 5.1

Problem solving generally is approached in a trial-and-error fashion. Why is not a more direct approach to problem solving adopted?

5.2.2 Planning

Generally, planning over levels is practiced in conjunction with an iterative-analysis mode of attack, primarily because the synthesis version is too difficult.

The iterative-analysis approach follows Figure 5.2, showing the feedback loops, and in particular the feedback loops between levels.

The analysis will generally involve network analysis and the presentation of the analysis results through time-scaled networks, bar charts, cumulative production plots, and so on.

Controls selected depend on the planner's experience and expertise, past projects, knowledge of the industry, as well as the situation and some creativity. Some solutions may be better than others. Generally, planners are after a satisfactory solution rather than some theoretically optimum solution.

If one approach does not work or does not work completely, then other approaches may be tried. Approaches are only limited by the planner's imagination.

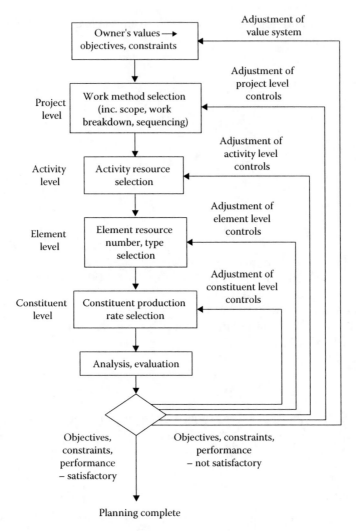

FIGURE 5.2
Iterative-analysis approach to planning over levels (definitional step not shown). (From Carmichael, D.G., *Project Planning, and Control*, Taylor & Francis, London, U.K., 2006.)

Whether any of these approaches are possible will be determined by the project and the nature of the project work.

Any control adjustment may, among other things,

- Lead to new cumulative resource (money) plots
- Alter the critical paths
- Lead to a rescheduling of activities

Values adjustment may redefine the planning or, for example, lead to the removal of physical, technical, or logic constraints affecting work.

5.2.3 Systems Engineering

Design in the systems engineering sense of Hall (1962) is promoted in terms of the following steps:

- Problem definition
- Value system definition
- System generation
- System evaluation
- Selection
- Action

A commonality with the previous set of synthesis via iterative-analysis steps is apparent. Such a systematic approach generates clear thinking at each step and lends objectivity to a procedure, which would otherwise be considered intuitive (Carmichael, 1981).

The start of the process depends on a postulated system extrapolated from experience or based on an idea. The end of the process depends on satisfaction of some desired and defined performance.

For many systems, the step of system generation is routinely obvious with the subsystem interrelations predefined by a body of knowledge in that system's discipline. Systems are commonly divided into lower-level subsystems ("subsystem delineation") to produce a tractable model and a tractable synthesis at a lower level. This is complicated by there existing an interdependence of each level, requiring knowledge at a higher level for synthesis at a lower-level subsystem. No isolated systems exist.

Example

When designing a structural frame, the subsystems are the beams and columns, while the subsystem interaction is in terms of equilibrium and compatibility relationships. The subsystem model is the (constitutive) relationship between forces/moments and displacements/rotations.

Exercise 5.2

Subsystem design is easier than system design and hence may have greater appeal. How do you guard against the possibility that the resulting collection of suboptimal (subsystem) solutions is not grossly suboptimal with respect to the overall system?

5.2.4 Direct Synthesis

The iterative nature may be removed (partly or fully) from the synthesis, if the system is synthesized directly. The difference between the analysis-based and direct procedures is in the quantity of *a priori* information about the system assumed. Analysis-based approaches assume a total system *ab initio*, while the controls (or system) emerge from direct synthesis. Extreme generality may be attained in the synthesis if no *a priori* knowledge of the emerging system is assumed. However, for a practical solution, certain leading aspects of the system are best assumed. The choice of how much system information to assume a priori is also a balance between the engineer's expertise and the computations involved. Synthesis can only proceed where some aspect of the system, usually the controls, remains free and adjustable.

5.2.5 Theory of Optimal Control Systems

Generally an engineer desires to synthesize a system, which is optimal in a certain sense; an objective resulting from an imposed value statement is implied. The optimum form is of central concern in optimal/optimum control systems theory, something that exploits the synthesis configuration and develops laws for control systems. The concepts of this theory are found extremely useful in synthesis generally. The concepts rest on very broad grounds, typical of techniques in systems theory, only conversing in the entities state and control (and sometimes output), which take on very definite meanings.

In simple terms, synthesis is equivalent to choosing the system controls; with an objective present, the optimum form selects the controls so as to extremize this objective. In addition, supplementary constraints are also usually present.

Optimum control theory in some engineering branches has elevated synthesis to a status approaching a systematic and exact science. This has occurred despite the ever-present, yet necessary expertise-based judgments, which recognize the existence, in all forms of synthesis, of certain intangible quantities that defy precise mathematical statements.

Exercise 5.3

How sure can a designer be when adopting a trial-and-error approach that the final solution is optimal or close to optimal? How can the designer be sure that the best solution has not been missed?

Consider mankind's established practices. Could you argue that, over time for each practice, mankind has shaken down to or reached something close to optimal by trial-and-error, even though at no point in time may anyone have explicitly stated an objective function (or even been aware of such things called objective functions) and carried out an evaluation based on it? This implies that the objective functions are in the background (that is, unstated or not explicitly acknowledged) and are implicitly guiding evolution of the practice.

Example: Trial-and-Error

Trial-and-error can be found wanting. Consider a project to resettle people who were living in an area to be flooded due to the construction of a dam. The project was responsible for ensuring the timely relocation of these people and ensuring that the people would reach a target income level within 3 years of resettlement. A number of livelihood programs were set up and regular socioeconomic monitoring of living standards was undertaken. After reviewing the results of the monitoring, the project team would adjust the livelihood programs to try to improve the people's income in order to reach the income target. This process went on for a number of years. This trial-and-error process turned out to be very expensive for the project. In hindsight, it was considered that a more direct synthesis approach would have been more appropriate.

5.3 Optimal Form of Synthesis

The optimal form of synthesis is most readily explained in control terminology and within the framework of optimal/optimum control theory.

Developing an optimal system involves finding admissible controls while minimizing/maximizing some objective. The objective is expressed mathematically as a function with, in general, both state and control variable arguments. Certain physical, economic, operational, engineering, and other constraints may be present, restricting the control choice. This choice may also be simplified if the search is confined to certain classes of systems. That is, system controls are chosen such that the system operates in some best way (according to the objective), while observing the constraints present.

Example

When designing a transportation system, by restricting the search to certain classes of vehicles, for example, train, motor vehicle, bicycle, …, the design is simplified over considering all possible types of vehicles.

5.3.1 (Optimal) Synthesis Components

A formulation for the optimal form of synthesis has the following components:

1. *A system model*
   This is often a constitutive equation, together where applicable with boundary and terminal conditions, ideally expressed in a standard form. It characterizes the system and enables the effect of alternative controls on the system to be predicted.

2. *Constraints*
   Constraints limit the range of permissible solutions and fix many of the system properties. (Boundary and terminal conditions might alternatively be regarded as constraints.)
   (1) and (2) are combined in some texts where they are broadly grouped as constraints. Where possible, in order to maintain understanding, this practice is not adopted in this book.

3. *Objective*
   The objective is derived from a value statement of the designer, planner, …, and is used to evaluate possible alternative controls. The controls that give the least or greatest value of this objective are desired. In general, the objective will be a function of both state and control variables, and will be a scalar quantity.
   Note that the term "objective" is not being used here in the lay or dictionary sense.

Exercise 5.4

On what basis can it be claimed that there are three components to (optimal) design and/or optimization?

Solutions are said to be *admissible* if the system model and permissible bounds, as defined by the constraints, are satisfied. Where a range of admissible solutions exists, the formulation might be referred to as being well posed. The objective provides the means by which the optimal controls are chosen from the set of admissible controls, in some best way.

Using a systems—state and control—foundation, superficially different synthesis applications can be shown to share a common mathematical basis, leading to common mathematical approaches.

Exercise 5.5

What value do you see in developing a common formulation for the synthesis configuration?

The following outline of the optimal form of synthesis is primarily for the single level, single objective case. Comments on the multilevel and multi-objective cases are given following the initial single-level, single-objective presentation.

5.3.2 Constraints

Constraints influence the result by isolating admissible controls, from all possible controls, and give sensible meaning to the result. Constraints may be defined on certain subsets, boundaries, or throughout the space and time domain of the system and are given in the form of inequalities or equalities.

In a sense, system models in the form of equations may be regarded as equality constraints over the space and time domain of the system. The system is constrained to belong to the class of systems whose equations are of this form. Terminal and boundary conditions may be likewise treated as equality constraints.

Constraints typically restrict the values taken by the state and control variables. The range of possible values that the states and controls may assume is reduced to a set of admissible values.

Exercise 5.6

For design practices, with which you have familiarity, what are the typical constraints?

5.3.3 Objective

An objective provides the means of quantitatively evaluating alternative admissible controls. The result is only optimal in the sense of the objective, which follows from value judgments, although computational tractability reasons may warrant introducing an alternative simpler objective. The latter objective obviously leads to suboptimal synthesis with respect to the original objective.

The terms "optimal" and "optimum" only have meaning in the sense of an associated objective. To use the terms "optimal" and "optimum" without reference to an objective has no meaning. Nevertheless you will hear people use the terms, particularly in managerial parlance ("management-speak"), without understanding this point; such usage is usually to impress rather than inform. Optimality implies an extremization (minimization or maximization) requirement on this measure called the objective.

The objective may be thought of as assigning a number to each admissible solution. The objective may be viewed as a function in which the controls play the role of the independent variables. The best value of this objective is sought.

The implementation of the optimal controls may not be possible for engineering, economic, or other reasons (that is, other constraints not allowed for in the mathematics of the synthesis). Knowing the optimal controls enables the implementation of a suboptimal version with a full understanding of the consequences of such action. In this sense, the optimal controls serve as a standard by which alternative controls may be compared.

Without loss of generality, minimization is commonly implied in all optimal synthesis studies. It will be appreciated that any formulation in maximization may be conveniently treated as one in minimization by means of a suitable negative transformation. Letting J be the objective,

$$\max(-J) = -\min(J).$$

Exercise 5.7

For design practices, with which you have familiarity, what are the typical objectives?

5.3.4 Probabilistic Systems

For states and controls that are random variables, the objective is now a random quantity, and hence an unsuitable measure. A suitable deterministic measure, over which the minimization may be carried out, is the expected value or first moment (in a probabilistic sense) of J, denoted $E[J]$. The expectation operation may be visualized as taking the average of the objective calculated for each of the possible values of its arguments. The variance, $Var[J]$, may also be incorporated into the objective function.

In certain applications, a measure or index of reliability may be relevant; that is, extremizing the index may relate to minimizing the probability of the system exceeding (both positive and negative senses implied together or singly) a particular limit state, or maximizing the probability of nonexceedance in order that the system attains a maximum level of reliability. Notice that this is a different situation to the one in which a system is designed for a given reliability (the probability of the state exceeding/not exceeding a given limit state is prescribed). Reliability in this context is a constraint.

Exercise 5.8

The world is inherently probabilistic. Why then do we always try to work with the deterministic version?

5.3.5 Multiple Objectives

The presence of more than one objective (resulting from multiple requirements) in general leads to different results (values of control variables) for each objective taken singly. In general, the results do not coincide, and hence the existence of more than one objective simultaneously poses additional questions. Usually the multi-objective case is resolved subjectively (and will vary from person to person), whereas the single-objective case can be resolved nonsubjectively (all persons come to the same result). Subjectivity is introduced in the three broad ways of addressing the multi-objective case:

- Through combining all objectives into a single objective (using, for example, Lagrange multipliers or weighting functions)
- Through converting all objectives except one to constraints
- By trade-offs or adjustments between the results for each objective considered singly

5.3.6 Alternative Terminology

Some writers refer to the collective of these three components (system model, constraints, and objective) as a "model of the optimization problem" or a "model of the decision making process"; such use of terminology only confuses and is not followed here.

The term *restraint* is sometimes used by writers instead of constraint.

The term "objective" is perhaps the most popular to denote that entity by which the best solution is chosen. However, different disciplines may use different terms such as *optimality criterion, performance index, performance measure, payoff function, figure of merit, merit function, goal, cost function, design index, target function, criterion, mission,* and *aim* (Carmichael, 1981). Where the objective is expressed mathematically, it may be referred to as an objective function.

5.4 Design Examples

Two examples are presented to indicate the synthesis formulation for design.

5.4.1 Structural Design

Consider the design of a framed building such as Figure 5.3a. The frame is made up of beams and columns (members) (Figure 5.3b).

In the design of such a frame, the designer selects

- The geometry of the frame—beam spans and column or floor heights
- Cross-sectional dimensions (widths and depths) of beams and columns (member)
- Material properties

All of these entities may be termed control/decision/design variables.

In selecting values for these controls, the designer is trying to end up with a frame with behavior that is acceptable in terms of

- Deflections of beams
- Sideways sway of columns
- Internal stresses/strains within the beams and columns

All of these entities may be termed behavior/state variables.

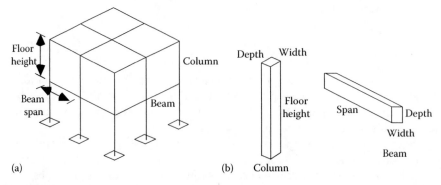

(a)   (b)   Column

FIGURE 5.3
(a) Framed building and (b) members.

As well, the designer is possibly trying to end up with a frame that is low in cost, which could generally be expressed as a function of the earlier controls.

The system model used in the iterative-analysis approach is the equations of structural mechanics developed in structural engineering textbooks. These are relationships between the control and state variables.

There are constraints that the designer has to be aware of, for example,

- Allowable maximum deflections
- Allowable maximum stresses/strains
- Constructability of the frame

The process may be firstly approached at the system level (that is, the frame level) and then at the subsystem level (that is, the beam and column level) and then at the sub-subsystem level (that is, the member cross-section level).

The process can be simplified, for example, by assuming certain materials such as steel or reinforced concrete. This converts such controls (the materials) to being constants in the formulation.

As (optimal) synthesis, the components are

Model: equations of structural mechanics

Constraints: on deflections, stresses, etc.

Objective: (minimum) cost

These are all expressed in terms of the earlier control and state variables.

The controls are chosen to satisfy the model and constraints while extremizing (minimizing) the objective. See Carmichael (1981).

5.4.2 Earthmoving Operation

Consider an earthmoving operation involving excavators and trucks moving earth continuously from a cut area to a fill area (Figure 5.4a).

The designer of this operation selects (control/decision variables)

- The number and type of excavators
- The number and type of trucks
- The haul road characteristics (length, condition, grades, ...)

These are the control variables.

In selecting these controls, the designer is trying to achieve for the operation

- Appropriate utilization (proportion of time working) of excavators
- Appropriate utilization of trucks
- A desired output or production

These are the state variables.

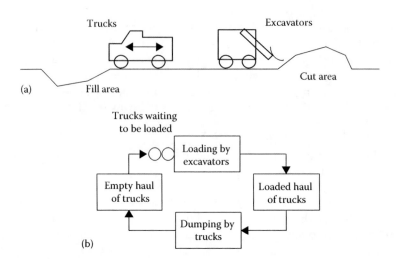

FIGURE 5.4
Earthmoving operation. (a) Pictorial and (b) block diagram.

As well, the designer may be trying to achieve least cost per unit production or least emissions per unit production.

For such operations, a suitable system model would be a queuing theory model or a simulation model. Subsystem models may relate to excavator loading characteristics and truck travel characteristics.

Constraints may relate to production levels, durations, and so on.

The designer guesses, or uses experience or industry knowledge to establish numbers and types of equipment, and haul roads. The operation is analyzed, and utilizations, production, costs, and emissions are obtained. The number and type of equipment and haul road characteristics are adjusted until the desired operation is obtained.

The designer is attempting to come up with an equipment and haul arrangement that is least cost per production or least emissions per production while satisfying any constraints present.

The iterative process may be simplified by fixing the type of equipment or haul road, or the designer may be constrained to using available pieces of equipment.

In terms of (optimal) synthesis, the components are

Model: queuing or simulation model, loading and travel relationships

Constraints: on time and production

Objective: (minimum) cost per production, (minimum) emissions per production

The controls are selected to minimize the objective while satisfying the model and constraints. See Carmichael (1987).

Exercise 5.9

Engineering operations are typically and historically arranged to give minimum cost per production (or cost per output). The objective function is one of (minimum) cost/production. The control/decision variables are variously equipment capacity and number, and the layout of the operation.

Now consider that you are interested in carbon emissions (resulting from the equipment) as well as cost, and consider any engineering operation that you are familiar with. For this operation, identify the control/decision variables, a suitable model of the operation, the objective function, and constraints for the minimum emissions/ production case.

Intuitively, how do you think the solutions to the minimum cost/ production and minimum emissions/production might relate to each other? Are they competing? Are they compatible? Are they the same?

5.5 Optimization Techniques

Optimization or optimal synthesis attempts to find the extremum (maximum or minimum) of the objective while satisfying any constraints and the system model. It may be approached through consideration of the mathematics alone, by ignoring sophisticated mathematics and doing the calculations numerically, by nonrigorous mathematics, by "gut feel" considerations, or combinations of these.

Where a mathematical technique is used, the particular technique is chosen to match the mathematical form of the synthesis components, whether they be algebraic, differential equations (and integrals), difference equations (and summations), or other.

5.5.1 "Gut Feel" Considerations

If the synthesis is sensibly posed and relates to a discipline in which the designer, planner, ..., has expertise, frequently the optimum or near optimum can be obtained by "gut feel" considerations.

Much of the literature on mathematical optimization exists because the underlying components are not sensibly posed. Well-posed, physically sensible formulations do not need intricate or sophisticated mathematics. There are many publications on mathematical optimization where mathematics-for-mathematics sake exists; generally, such publications give little time to the formulation, but rather jump straight into deriving results to silly formulations.

5.5.2 Numerical Approaches

Numerical approaches simply calculate the objective for a range of values of its arguments. That set of values that gives the lowest (or highest) number for the objective is taken as the optimum. Spreadsheets are particularly useful for such approaches and are liked by practitioners.

5.5.3 Calculus

There exists the calculus of extrema for finding minima, maxima, and points of inflection of functions. These points are obtained by setting the first derivative of the objective function to zero (necessary conditions for extrema). Checking the sign of the second derivative of the function establishes the type of extrema found.

Extensions using Lagrange multipliers attempt to deal with the presence of constraints.

5.5.4 Pontryagin's Maximum Principle

A generalization of calculus and Lagrange multipliers to the dynamic system model case leads to Pontryagin's maximum principle. The resulting necessary conditions of optimality are a set of simultaneous equations in the state and costate (Lagrange multipliers) variables.

5.5.5 Dynamic Programming

Similar dynamic cases may be handled using dynamic programming. The technique also extends to sequential or staged systems. Dynamic programming is based on Bellman's principle of optimality stated in the 1950s:

> An optimal policy has the property that whatever the initial state and initial decision are, the remaining decisions must constitute an optimal policy with regard to the state resulting from the first decision.

5.5.6 Mathematical Programming

Where the synthesis components are a collection of algebraic relationships, mathematical programming can be applied.

Linear programming gives the result where the relationships are linear.

Nonlinear programming techniques apply where the relationships are nonlinear. Innumerable methods have been proposed and include geometric programming; quadratic programming; search methods; gradient methods, function approximation; penalty function methods; and methods of feasible directions.

To demonstrate their optimality, sometimes reference is made to the Kuhn–Tucker conditions, which are necessary for an optimum for the constrained optimization case.

Exercise 5.10

Itemize the approaches to optimization that you have come across and what synthesis formulation was being attempted. At the least, you should have come across linear programming and calculus in high school.

5.6 Project Planning

> If you don't know where you are going, you will end up somewhere else.
>
> **L. Peter**
>
> If you don't know where you are going, any road will take you there.
>
> **Anon.**

Carmichael (2006) explores planning in multilevel synthesis terms. Generally, other publications are unaware of the synthetic nature of planning (though many have converted resource leveling and resource-constrained scheduling, and the attainment of the shortest network path into optimization terms, but that is not the same).

Planning establishes how and what work will be carried out, in what order and when, and with what resources (type, and number or quantity, additionally expressed in a money unit). "Resources," in a planning sense, refers to people and equipment only. To plan (verb) is the act of choosing the controls (method, resources, and resource production rates) throughout the project duration. A plan (noun) is the outcome of planning.

5.6.1 As Synthesis

Planning, as mentioned, is an example of synthesis. As such, there are multiple results (choices of controls) possible. In most cases, planners are only after a satisfactory solution, or a result that they can live with, and do not put additional effort into searching for the optimal result. A planner may also be under pressure to come up with quick results.

Most planners are unaware of and do not understand the components of synthesis, and so they never know where they are relative to the optimum. They are unable to vocalize or formulate the synthesis components. Such discussion goes to the very heart of current planning knowledge being in its infancy, and current planning education (read training) being superficial and low standard.

5.6.2 As Iterative Analysis

Alternatives might be arrived at through the generation techniques used in a value management/"analysis"/engineering study, which aims at developing alternatives that perform a desired function but at a lower cost, or in a work study, which is concerned with the method and timing of work activities and their optimization, or in creativity studies. Both value management and work study are no more than synthesis, approached in an iterative-analysis fashion, themselves.

In the common iterative-analysis approach to planning, plans develop as thinking on a project progresses. Often, during the initial or early phase of a project, a high-level plan is developed. The key milestones might be identified, and some level of guesstimation is used on the timeline needed to achieve the milestones. The true scale and complexity of the scope may not always be clear, and some assumptions have to be made. Also it is known that resources will be needed, but their commitment or availability for the project may be uncertain. As time moves on, the plan is continuously fine-tuned by including changes to the sequence of work, time frames, addition/removal of resources, etc.

Exercise 5.11

Even though planning is approached in an iterative-analysis fashion by practitioners, why are only zero or a few iterations common? The exercise is talking about planning (before the work starts), not replanning (after work starts), or planning where the scope or objectives may change.

5.6.3 Project and End-Product

A project comes about because of an identified need or want for some product, facility, asset, service, etc. This end-product is achieved through a project.

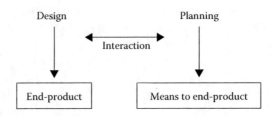

FIGURE 5.5
Fundamental project synthesis undertakings.

This distinction sometimes causes people confusion, and many people are not aware of the distinction, or of the need to make a distinction. For example, people sometimes refer to a building as a project. It is not. The processes that go together to materialize the building are the project. The building is the end-product. (It is acknowledged that the definition of a project is sufficiently flexible to include the operation and maintenance phase of a product within what is called the project. However, this is not the issue here.)

It is thus important on projects to differentiate the "end-product" of a project from the "means to the end-product" or the project itself.

There are two synthesis formulations involved on projects, namely, that related to (Figure 5.5)

- The "end-product" (*design*)
- The "means to the end-product" or the "project" (*planning*)

The materialization of a project's end-product is nonunique and can be performed, possibly, in an infinite number of ways. In a similar way, the selection of the end-product is nonunique, and there are possibly an infinite number of versions of end-products.

There are correspondingly two sets of objectives and two sets of constraints:

- End-product objectives and constraints
- Project objectives and constraints

These define two synthesis formulations. Some project objectives and constraints may reflect or be influenced by end-product objectives and constraints.

The preferred materialization of the end-product is selected based on the project objectives and constraints. Work method, scope, resource, and resource production rate considerations on projects follow.

The preferred end-product is selected based on the end-product objectives and constraints. Form, geometry, function, materials, finishes, appearance, etc., of the end-product follow.

(Note that the design process itself can be interpreted as a subproject—which in turn can be planned, but this is not what is meant here; the earlier reference to design is to the design function.)

Exercise 5.12

Lecture timetables, transport timetables, television programs, training schedules, and bar charts are all examples of ways the results of planning are conveyed. Suggest what the underlying or background synthesis formulations might have been.

Exercise 5.13

Consider a typical project. Planning might be undertaken by various project participants—owner, consultants, contractor, subcontractors, and suppliers. How might the synthesis formulation of each differ?

5.6.4 Conventional Thinking on Planning

Existing publications on planning, with the exception of Carmichael (2006) do not appreciate that planning belongs to the synthesis configuration. User-friendly computer packages are available to assist the planning process, but these only reinforce an analysis view of planning. The challenge is to understand what planning actually is; people are going through the motions of planning but do not understand the fundamentals of their trade.

The fundamentals of planning are most easily seen if a systems approach is adopted, with control systems terminology preferred for its precision and usefulness over conventional planning terminology. In particular, state space and related terminology of modern control theory are useful.

Existing treatments on planning jump head first into network analysis, both deterministic (CPM—critical path method) and probabilistic (for example, PERT—program evaluation and review technique, and Monte Carlo simulation), with all refinements including overlapping relationships, and the use of industry-preferred packaged software. This uninspired and pedestrian approach is justified falsely by the reason of "practicality."

A more inspired approach is possible, if it is acknowledged that planning is synthesis. As such, the results of planning are not unique.

In most cases, planners are only after a satisfactory result, or a result that they can live with, and do not spend the additional time searching for the optimal result. A planner may also be under time pressures to come up with quick results.

However, planners expediently reverse the logic and deflect attention from their inability to come up with best results, on time pressures and pseudo "practicality" arguments, when in fact most planners do not understand the synthesis nature of their job. Planners are unaware of and do not understand the components of synthesis, and so they never know where they are relative to the optimum. Designers are similarly guilty.

5.6.5 Planning Function

Planning establishes the value of *controls* (decision, action, ...), throughout the project duration. The controls contain information on method (including sequence), resources (type, and number or quantity) (and hence money), and resource production rates (or equivalent). The project can be viewed as a dynamic system.

A hierarchical and staged approach to planning might be adopted. Four main hierarchical levels might be considered—project, activity, element, and constituent. Decoupling the hierarchical levels and stages facilitates the solution, but introduces iterations. Determinism would generally also be preferred by planners.

Planning controls are selected over the full-time frame of a project for a globally optimum solution. Planning separately on a stage-by-stage or even day-by-day basis is suboptimal. Control selections early in a project constrain future updated control selections.

Frequently, a dominating constraint is that of a desired project performance or behavior/state over the project duration; such a constraint may eliminate most of the potential for optimization. Control selection implies knowledge of work practices and (fore)thought as to what might happen later in time. In this last sense, planning is said to be proactive; the alternative is to be reactive and not carry out forethought as to what might happen. The state contains information on cumulative resource usage, cumulative money usage, or cumulative production or equivalent. Project output is also an indicator of project performance, but for the stripped-back project formulations usually used, output and state are the same.

5.6.6 Planning Components

Constraints: All projects have constraints such as funding, resourcing (people and equipment), environmental (natural), and political constraints. As with objectives, many people confuse a project constraint with an end-product constraint. End-product constraints may influence project constraints.

Constraints may be stated at all project levels.

Objectives: Commonly, project objectives say something about project cost, project duration, and deviation from specification, but other objectives are possible. And these may apply throughout the project—for example, (minimum) deviation from specification—or at the final (terminal) point of the project—for example, (minimum) duration, which is related to the project completion time. That is, general project objectives will contain a component over the time domain of the project and a component at the final (terminal) point.

Project objectives may be expressed at the various project, activity, element, and constituent levels.

End-product and project objectives may derive from higher-level values within an organization, for example, originating from a corporate idea. Such values may also reflect political, marketing, environmental (natural), ... concerns.

End-product objectives and project objectives (and constraints) may relate, for example, to

- Money (end-product—sales, benefit:cost ratio (BCR), present worth, ...; project—cost, budget, ...)
- Duration (end-product—lifetime, ...; project—duration, ...)
- Resource usage
- Quality issues
- Community acceptance
- Environmental (natural) effects
- Safety
- Risks
- Public impact
- Extreme event impact (floods, cyclones, ...)
- Social impacts
- Geotechnical considerations

These are expressed in terms of end-product matters or project matters, as the case may be.

Exercise 5.14

What is planning to you? Do you see both design and planning as belonging to the synthesis configuration?

Example

Planning establishes how work will be done, in what order, and with what resources.

Consider the simpler situation where it has already been largely established how the work will be done. As such, the controls available to the planner are the order or sequence of the work including when particular work items will be done.

This establishes when resources (people, equipment) are used (and hence when money will be used).

What the planner is trying to influence (state variables) is the duration of the work and the utilization of the resources.

There may be constraints on costs, duration of the work, and resource availability.

Ultimately, what the planner might be trying to achieve is a work sequence of minimum cost or minimum duration or a compromise between these two.

A suitable system model is the time-scaled critical path network or linked bar chart, with connections to resource histograms and cash flow diagrams.

As the planner adjusts the sequence of work, so alternative cash flows, resource utilizations, and work durations evolve. By a series of iterations, the planner settles on one particular sequence of work.

Using a synthesis approach, the components to planning are

Model: time-scaled network or linked bar chart
Constraints: on work duration, costs, resource availability
Objective: (minimum) cost and/or time

5.6.7 Project "Control"

What is popularly called "project control" is not control in the sense used in this book. Rather, it is a lay usage of the term; it refers more to "containment" or "restricting" cost and time, rather than getting the project to where the planner wants it to be. A more accurate name for what people call "project control" is replanning, that is, planning carried out later in a project.

What replanning, when necessary, does is come up with revised values of control variables, for the remainder of the project. The nature of the control variables is unchanged and contains information on method, resources, and resource production rates.

Replanning does what is essentially embodied in Bellman's principle of optimality, which was devised for different purposes. The principle states: *An optimal policy has the property that whatever the initial state and initial decision, the remaining decisions must constitute an optimal policy with regard to the state resulting from the first decision.* The words "decision" and "control" may be interchanged. See Carmichael (1981).

In essence, replanning is no different to planning. It is planning carried out subsequent to any original planning, using updated initial conditions, shortened time horizon, updated constraints, and possibly updated objectives. As such, most of any discussion on planning applies to replanning; there is no need to even mention a term called "project control."

Project performance (the project state) is monitored on an ongoing (usually regular) basis, *compared* with planned performance (typically in a report) and, if necessary, *corrective action* is taken to send the project in a desired direction (on a desired state trajectory) (Figure 5.6). The controls bringing about corrections may be in the form of method (including sequence), resources (or money), and/or resource production rates (or equivalent) input to the project. The controls are chosen to extremize the objectives while satisfying any constraints present. *Applying (changed or different) controls amounts to replanning.* Accordingly, planning (in the form of replanning) might be referred to as an ongoing or evolutionary process. The "planning phase" of a project occupies the total project duration; hence, for people to talk of a "planning phase" is incorrect.

Generally, it can be said that prior work done constrains future (updated) planning actions. Hence, the constraints are changing as a project progresses. The flexibility in choice of controls decreases with time.

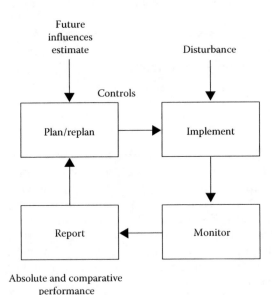

FIGURE 5.6
The replanning process.

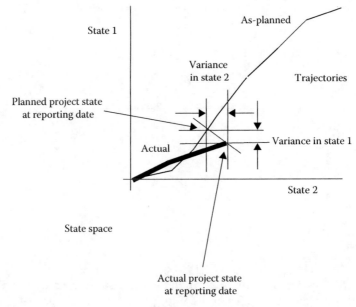

FIGURE 5.7
Generic way of representing project variances in a state space.

One useful report is that which shows together all *variances* (not the statistical variance), that is, the difference between as-planned performance and actual performance for all indicators of project performance (that is, state). A generic form of collective variance reporting is shown in Figure 5.7. Figure 5.7 shows a two-dimensional *state space* (a space with state variables as axes), but the same holds true for where the performance of the project is reflected in more than two states; the space becomes a hyperspace for dimensions greater than three. (For dimensions greater than three, tabular reporting of these variances or pairwise reporting of these variances might be preferred.) Two state trajectories are shown in Figure 5.7—one corresponding to the as-planned behavior and the other corresponding to actual behavior. The variances in the states are also shown.

As an example, the two states in Figure 5.7 might be cumulative resource usage and cumulative production. The variance in cumulative production will indicate whether a project is "ahead/on/behind schedule"; production performance not only reflects actual resource production rates, but may also reflect delays, and popular reporting may be more accustomed to schedule and delay reporting rather than reporting as in Figure 5.7. The variance in cumulative resource usage, when converted to a money unit, will indicate whether a project is "over/under budget." Popular reporting may also be more accustomed to cost reporting rather than reporting as in Figure 5.7; in that sense, it might be preferred to aggregate the cumulative resource usage states across all resource types and express this as a cumulative cost state.

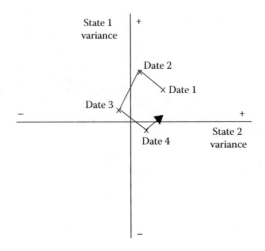

FIGURE 5.8
Progressive plot of state variances.

The variances from Figure 5.7 might then be summarized over time in diagrams such as Figures 5.8 and 5.9 to give a historical record of project performance.

A way of reporting variances, less useful (but one that is conventionally done) than Figure 5.7, is shown in Figure 5.10, where individual states are plotted against time. This only gives a part picture of what has happened on the project up to the reporting date. For example, where cumulative resource usage is the state being represented in Figure 5.10, the figure may indicate that less resource numbers have been used than had been planned, but says nothing as to whether production is as-planned, ahead of as-planned, or behind as-planned.

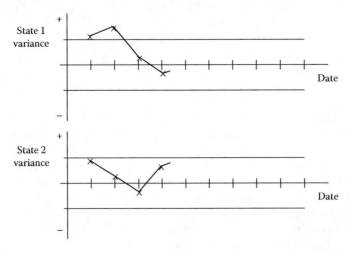

FIGURE 5.9
Separate plots of state variances.

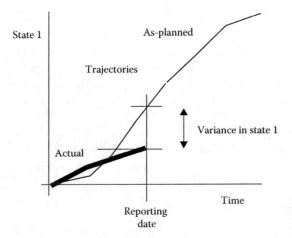

FIGURE 5.10
A way of showing project variances less useful than Figure 5.7.

Earned value thinking introduces a modification to Figure 5.10 when money and production/schedule are being looked at together, but still Figure 5.7 is a preferred form of representation and removes all the limitations inherent in earned value reporting. The limitations in the earned value approach stem from time being used as the independent variable in Figure 5.10; if states are used as the coordinates as in Figure 5.7, all the limitations disappear. One day, industry might take this on board.

Exercise 5.15

List possible sources of control that could be used in the following situations in order that the project does not diverge further from the planned state. No change in scope is allowed, and the specification cannot be altered. No unethical practices are allowed. It is assumed that the original planning and design for the project have been undertaken in an optimum fashion, and there is no "fat" available for trimming.

- The project is running behind production/schedule.
- The project is going over budget.

Exercise 5.16

Project closeouts commonly have a heavy emphasis on a fixed completion date. Planning may be done backward from this point. What special planning issues would this create compared to planning being done in a forward sense?

Exercise 5.17

Some people say that they plan backward from a completion date, but actually plan forward and adjust iteratively if the completion date exceeds whatever is desired. Some people use a mixture of going backward and going forward. Some people go backward. Give comment on the efficiency of these different approaches. Would a combination of both forward and backward planning reveal a planning solution better than either forward planning or backward planning alone?

Exercise 5.18

Most people (and textbooks and computer packages) talk of "project control" or similar terms, and think that it is something magically different to planning. In essence, what people are doing when they are doing "project control" is replanning (which is itself planning). The planning (and replanning) formulation fundamentally includes constraints, and this deals with the lay usage of the word "control." Why do people regard "project control" as something different to planning? Is it because they have never thought about the synthetic nature of what they are doing and are blindly following the herd?

There are many people who attend seminars on planning, and who think that learning about "project control" will make all their workplace troubles go away. When told that it embodies no more than monitoring/reporting/replanning, and can be learned in a very short time, and that if they already know planning, then there is little new to be learned, you sense their disappointment. They want the magic that is embodied in the magical term called "project control." They want a panacea for all their project troubles. Why do people think like that?

What does "pest control" mean? Is it using the term "control" in the sense of this book?

5.6.8 Financial Planning

Financial planning refers to all that involved in deciding the usage of money on a project. The term *budgeting* may also be used to mean financial planning, and the term *budget* to mean a financial plan.

Financial planning is a sub-issue of overall planning, where emphasis is placed on the different level controls that can influence money usage (including recoupment), and all other controls are held fixed. Resource usage is converted to a unit of money.

Budgeting for projects, some might argue, is a more difficult task than budgeting for a business. Projects are unique and have to be developed from first principles each time, compared with a business where next year's operations could be expected to bear some similarity with this year's operations. However, budgeting in a business via work packages (and later monitoring against these work packages) is little different to that involved in budgeting for projects.

5.6.9 Human Resource Planning

Human resource planning refers to all that involved in deciding the usage of people on a project.

Human resource planning is a sub-issue of overall planning, where emphasis is placed on the different level controls that can influence the usage of people resources, and all other controls are held fixed.

5.6.10 Resource Planning

Resource planning refers to that involved in deciding the usage of resources (people and equipment) on a project.

Resource planning is a sub-issue of overall planning, where emphasis is placed on the different level controls that can influence resource usage, and all other controls are held fixed.

5.6.11 Strategic Planning

Strategic planning is no different to project planning, only over possibly a longer time frame and with less definite data. The end-product might be called aspirations or similar. The planning establishes the value of *controls* (decision, action, ...) throughout the time frame considered. The controls contain information on method, resources, and so on.

5.7 Management

Unfortunately, the terms "management" and "manage" are used in the Lewis Carroll sense. There are multiple lay and dictionary uses of the terms, and often all meanings can be found to be used in any one document. No attempt at rigorous terminology use is attempted. Being called a manager sounds much better than being called a supervisor, administrator, clerk, or whatever the true job entails, and hence people called "managers" proliferate throughout the workforce. "Manage" might mean accomplishing or achieving something despite obstacles; handle; direct or control;

to establish authority, domination and discipline; and so on. Discarding these lay meanings and hence activities that fit these meanings, along with the activities of pseudo-managers, the core of management belongs to synthesis.

Project management and general management are very young by comparison to established engineering disciplines. This means that the models used in management are also very crude by comparison. Contributors to management theory development generally also come from nontechnical backgrounds, which means that their contributions are restricted to simple verbal models. True advancement in management will not happen until quantitative modeling becomes the norm.

Within project planning, the controls relate broadly to method and resources (people and equipment). The term "resources" might be used by most people in a lay sense to mean nearly anything (including "time"), but within planning, it very specifically refers to people and equipment only, and no more. Within management, the range of controls is very wide and includes anything that drives a company or project toward its desired output.

Management was first identified as belonging to a synthesis configuration in Carmichael (2004). No existing approaches to project management or management acknowledge that management belongs to a synthesis configuration, but this possibly reflects the crudeness of the state-of-the-art of management, and the deficient backgrounds of many teachers and researchers of management.

The development of management suffers because anyone, or even your pet, can call themselves a manager; the term is not protected. People are often given the grandiose title "… manager" by a company, more than likely as compensation for the company giving lower pay, when a more appropriate title should be used. Such people then have to get a quick-fix training course on how to be a manager, even though such people are not truly doing "management." This market is well-serviced by an oversupply of shallow seminars and paperback books based on opinion and anecdote, rather than truly nonsubjective information supported by the scientific method. There is also a healthy market promoting management "fads" but given grandiose names, a healthy recycling of existing practices but given new grandiose jargon, and a healthy population of people promoting themselves as management gurus.

Compared to managing a project, with its uniqueness, managing an established ongoing enterprise could expect to be characterized by

- More certainty over time
- Tasks may be repetitive
- Well-understood roles and interpersonal relationships
- Relatively stable work situation

For some work practices, it is possible to think of them as projects and apply project management principles. The term "management by projects" is sometimes used to describe such an approach and is sometimes abbreviated to "MBP."

Many people carry out this practice without adopting a formal term to describe the process or recognizing that in fact this is what is happening.

5.7.1 General Management

Consider a company (and its equivalent in the public sector, or generally an organization). Management may be viewed in the following synthesis terms.

State, performance, or output: The performance or output of a company can be in terms of profit, repeat business, customer goodwill, stakeholder satisfaction, and so on.

Controls: The means of achieving any desired performance is through the manager's selection of controls (which are quite extensive) and include staffing (number, type, payment, …), work arrangement (manner in which work is processed), procedures (rules of staff behavior), and so on.

Objective function: An objective function is required in order to reduce the number of admissible controls to a preferred control. Managers seem to be preoccupied with minimum cost, or maximum profit, and production, but broader triple bottom line (TBL) and corporate social responsibility (CSR) issues are starting to get a mention.

Model: Connecting the input and output is some model or representation of the organization (the system). Current model development of organizations and people is very crude, primarily because most researchers are constrained by their educational backgrounds to using verbal models, and using surveys and interviews as substitutes for research tools. The state-of-the-art of management will not progress until quantitative models are developed.

Conventional management texts, practices, and teaching do not realize that management belongs with synthesis. Once this is realized, the art of management can develop beyond its current primitive status. Without acknowledging its synthesis roots, the state-of-the-art of management will go nowhere.

This perspective is quite a different way of viewing management compared to the *classical* view of "planning," organizing, staffing, directing, "controlling," and coordinating, though within each of these function areas, synthesis formulations can be developed.

Sub-issues of general management include synthesis involving people, equipment, finance, etc., but these are not mutually exclusive sub-issues.

Dealing with the synthesis of these sub-issues in isolation will be easier than dealing with all sub-issues together, but this necessarily will lead to a collection of solutions suboptimal with respect to the whole.

5.7.2 Project Management

Project management is commonly described in terms of various management functions—management of scope, quality, "time," cost, risk, contract/procurement, human resources, communication, and possibly others. However, such checklists are not internally consistent and also duplicate aspects. Projects may also be viewed in a chronological sense, where a project is viewed as passing through phases.

The synthesis formulation for project management can be summarized in the following.

State, performance, or output: Project performance or output might be described in terms of resource (people, equipment, materials, …) usage (and hence costs), production (and hence schedule), quality, safety, and the like.

Controls: The controls that the project manager has at his/her disposal are many and include resource type and number, work method, procedures, and so on.

Objective function: Typically, projects are considered successful if cost, duration, and accidents are minimized, and quality maximized.

Model: The same comments and restrictions on modeling within general management apply to project management. The state-of-the-art of management modeling is very crude.

This thinking is quite different to conventional project management approaches.

Sub-issues of project management include synthesis involving people, equipment, finance, etc., but these are not mutually exclusive sub-issues. Dealing with the synthesis of these sub-issues in isolation will be easier than dealing with all sub-issues together, but this necessarily will lead to a collection of solutions suboptimal with respect to the whole.

5.7.2.1 *"Time" Management*

The term project "time management," that is, the management of anything involving time on a project, is commonly (and incorrectly) used. There also exists the term "personal time management" relating to how people "use their time." Neither form actually manages time, but the terminology is widespread. Also, some people loosely refer to planning and time management synonymously.

Strictly, time is the independent variable, and projects should be regarded as dynamic systems because of this.

Exercise 5.19

In many ways, the field of management is very crude when compared, for example, with established engineering disciplines of age where there is little chance for you to contribute anything of size. However, there is potential for someone to make a significant contribution to the state of the art of thinking about management.

Project management and general management are very young by comparison to established engineering disciplines.

Most of the contributors to general management seem to have come from nontechnical backgrounds. The important contributions to project management seem to have come from people with technical backgrounds.

This has led to general management models being primarily verbal models. However, models preferred by people with technical backgrounds are quantitative models and often mathematical in form. The imprecision of verbal models to a person with a technical background is always a concern.

In the distant future, management models will be on a par with engineering models. The argument advanced against this is that management is concerned partly with human behavior, and people are extremely fuzzy entities. The counter to this is that perhaps we are not looking at the modeling process in the right way—some lateral thinking is needed.

What are your views on these thoughts?

5.8 Risk Management

It is argued here that a rigorous technical definition of risk is needed in order to understand risk management and that lay and dictionary definitions should be banished from everyone's thinking.

Risk is the exposure (loss or gain) to the chance of occurrences of events adversely or favorably affecting the project/business/... (Carmichael, 2004). Risk is related to the probabilistic output of the project/business/... being examined, resulting from uncertain events.

Be aware though that most people (including standards, codes of practice, books, and papers) get the definition of risk wrong, typically by adopting lay or dictionary meanings, also use multiple meanings for the one word in the one document, and cannot agree on what risk is. Risk is one of those words encapsulated by the Lewis Carroll quote. Risk is not a chance, a probability, or a likelihood. Risk might be a possibility, but such a definition does not let you do anything with it. These are street-talk and dinner-party meanings and are totally unsuitable for a rigorous development of what risk management is.

Some of the confusion over the definition of risk can be seen through people using the terms "risk event" (source, factor) and "risk" interchangeably. Risk is the downstream exposure (output), originating from an upstream risk event/source/factor (input). A risk is not a risk event (an output is not an input). A risk arises as a result of a risk event. Most people confuse the two. This is particularly so in what is popularly termed "risk identification." What is meant by "risk identification" is usually "risk event identification." Certainly, potential consequences or outcomes will be considered, but only as an exercise in then working backward to establish the underlying risk event or source—the event that results in the consequence or outcome.

Another example of the confusion as to what risk is can be seen in the managerial speak of "risk versus return" or "risk-return." Here risk seems to be used in the lay or dictionary sense to mean probability; higher positive returns should coincide with lower probabilities of getting a positive return, and vice versa. However, a return's probability of occurrence, in conjunction with its magnitude, represents the risk.

Example: Project

On any project, there are many things that can happen that lead to cost overruns and time overruns (beyond that originally planned)—the exposure. These include material delivery delays, inclement weather, unavailable resources, accidents, and so on. These initiating events (input) are the risk events/sources, and their occurrence is uncertain. They lead to overruns, which themselves accordingly contain uncertainty. The risk is related to the probabilistic overrun (output).

Example: Accidents

Accidents—the exposure—in the workplace result from, among other things, events that are not anticipated. The occurrence of any initiating risk event (input) is uncertain. It follows that the occurrence of the accident (output) is also uncertain. The risk is related to the probabilistic occurrence and severity of the accident.

Exercise 5.20

Why is it that people can quite happily use the terms "risk" and "risk event" interchangeably when they clearly mean different things? How rigorous are you in your usage of terminology?

5.8.1 Risk Management Process

The usual form seen in texts covering risk management describes the risk management process in terms of steps such as

- Establishing the context
- Identifying (risk events)
- Analyzing the risk situation
- Establishing consequences/assessment
- Responding/treatment

There may be feedback between steps in attempting to refine the process or as secondary influences materialize. Often identification is intertwined with analysis and establishing consequences/assessment and the three proceed hand in hand.

For a changing situation, risk management becomes an ongoing exercise over time.

The risk management process is a particular case of synthesis via iterative analysis, though different terminology is used and the number of steps might be different.

Exercise 5.21

The risk management process represents a special case of synthesis via iterative analysis. How do such special processes come about? Why do not people start with the existing and more generic synthesis and either use this or specialize this? Or do people have trouble starting with the general and going to the specific; is it easier to approach any situation from the specific?

5.8.2 Definition

The definitional step in the risk management process includes establishing the context

- From whose point of view is the risk management being undertaken—client, consultant, contractor, public, ...
- The nature of the risk being studied—safety, cost, production/schedule, ...

The step also includes value judgment characterization of the risk or some measurement scale for risk (called risk assessment in risk management parlance). For example, high risk might correspond to a large exposure in dollars; consequences that impact humans may be regarded as primary risks, while consequences related to budget and operations may be regarded as secondary risks. The value judgments of the person or organization, for which the risk study is being undertaken, introduce subjectivity into the process. Different people and organizations characterize consequences, and hence risk, differently. People rank or prioritize risks differently.

The characterization of the risk or some measurement scale for risk is based on linking likelihood and consequence. Matrices, with axes of likelihood and consequence, and entries of risk levels, are commonly used. As the risk management process progresses, and highlighting the subjectivity involved, people interestingly might adjust or manipulate a likelihood or consequence to achieve a favorable answer, or to give a client what it wants.

Exercise 5.22

What happens if the risk management is being carried out from the point of view of more than one person or more than one organization? Effectively, you have more than one set of value judgments (the equivalent, in form but not in terminology usage, of the multi-objective case). How do you deal with this situation?

5.8.3 Objectives and Constraints Statement

Objectives and constraints exist as in any synthesis configuration. The objective provides the basis for the control or decision choices made (called responding or treatment in risk management parlance). Examples: (minimum) risk associated with cost overruns. The constraints restrict the control choice. For example, certain consequences or controls may be totally unacceptable.

Typically, risk is assumed in a downside sense, and hence minimization is usual if risk is part of the objective. However, a more general view is that risk can have a downside or an upside.

Objectives and constraints are selected from the point of view of the person or organization, for which the risk study is being undertaken. Different people and organizations will select different objectives and constraints.

Exercise 5.23

The objectives result from value judgments. How are these established for organizations, as opposed to individuals?

What happens, say, if there are two objectives—(minimum) risk and (minimum) cost. How is such a multi-objective case handled?

5.8.4 Alternatives Generation

Different scenarios might be envisaged for the risk management situation being looked at.

To start the iterations, a common approach is to assume that no controls are present, or the controls corresponding to the status quo are present. (This situation is then analyzed, evaluated, and the controls adjusted.)

5.8.5 Analysis

Analysis depends on firstly having identified all possible or all meaningful risk events/sources/factors. Identifying risk events can involve creative thinking and can use idea-generating techniques. The identification of risk events/sources/factors may be carried out with the help of checklists, interviews, brainstorming sessions or personal and corporate past experiences, and other more specialized approaches. In safety matters, this step might be referred to as hazard identification or developing a hazard scenario. The identification step is the Achilles heel of risk management; if an event is not identified, no subsequent control change takes place.

Exercise 5.24

As part of doing business with other countries, selection of the preferred currency of payment and sensitivity analysis to movements in the value of the currency are standard decisions and practices that are taken.

What is not usually considered is the impact of gross currency movements, such as has occurred in the recent past. Such movements in currency values appear not to have been able to be foreseen. That is, the risk event or source would not have been identified.

Where does this leave the practice of risk management, if the cause of a major risk is not identified?

What is termed "risk analysis" in risk management parlance involves the conversion of the risk event information into information on consequences or outcomes.

The analysis step may be carried out qualitatively or quantitatively, using whatever technique and approach is appropriate. Commonly, analysis may be carried out in two stages. The first stage is a qualitative analysis that carries through a subjective view of consequences. The second stage is a quantitative analysis that presents a more definite view. For cursory risk management, only the qualitative analysis may be carried out.

A quantitative analysis would generally employ some numerical approach, often using a computer. It revolves around establishing quantitative estimates for costs, times, ... and uncertainties. A prior qualitative analysis is recommended in order to obtain an overall understanding. Such an analysis may be carried out in order that low-impact consequences are excluded from a more detailed study. If no more is done, at least a qualitative analysis is considered essential.

Exercise 5.25

In what circumstances would you anticipate that a qualitative analysis by itself would not be sufficient? At what point does a quantitative analysis become essential?

Exercise 5.26

Think about previous projects or work matters with which you have been involved. List some of the things that went wrong or astray because the consequence was not foreseen in magnitude or type.

Interestingly, in analysis, many people assume that the probability or likelihood of a risk event carries through from event to consequence. This can only be assumed to occur if there is a direct relationship (one-to-one, linear) between event and consequence. However, the usual assumption made by people is that the probability of an event/source (for example, a hazard) is the same as the probability of the consequence (for example, injury).

5.8.6 Evaluation

The outcome is evaluated according to the stated objective. Commonly, low (minimum) risk or zero risk is desired.

5.8.7 Iteration Feedback

Having done the evaluation according to the objective, it may now be necessary to alter the controls. (In risk management parlance, this step is referred to as responding or treatment.)

Control selection might involve actions that change or remove aspects of the risk source and its likelihood of occurrence, and/or aspects of the consequence and its likelihood. Controls are only selected if they influence the value taken by the objective function. There is also the "do nothing" option; that is, no new controls are selected or no existing controls are changed. Low or minimal risks may be accepted without further consideration.

Two timings of the control actions may be recognized:

- Immediate responding. Actions are taken to address the risk issue now.
- Contingency responding. Actions are developed to deal with any possible adverse consequences, but are only implemented should an underlying risk source/event occur. (This might be referred to by some writers in a lay person's sense as a contingency plan, but such terminology is misleading.) (Note: The term "contingency," in projects, is also used to refer to a cost or time reserve.) Trade-offs may be considered in terms of responding at different points in time and the different consequences associated with responding at different times.

Control choices might be categorized according to whether they

- Eliminate (including remove, avoid, full-transfer) a risk (value of a (minimum) risk objective equals zero)
- Reduce (including part-transfer) a risk (lower value for a (minimum) risk objective)
- Leave unchanged (including accept, retain, assume) a risk (value of a (minimum) risk objective remains the same)

Example: Cost and Time Reserves

A cost reserve or a time reserve on a project may also be called by the name "contingency." It is intended to allow for KUKs (known unknowns). Reserves commonly occur in cost estimates. Float in project programs is a form of reserve. A tolerance or "space" in a performance specification is another example. Reserves are an example of reduction.

The reserve is the control used. It reduces the objective function—risk associated with project duration and cost overruns.

Typically, reserves are added as a percentage; for example, an extra 10% might be added to a project estimate. This percentage is based on whatever worked well last time, whatever is customary in that industry, or whatever seems right according to "gut feel." But the rationale of adding a percentage is not sound. Ideally, risk should be considered at the component level, and a contingency built up to the project level. But nobody seems to do this. Maybe one day industry will take this on board.

Example: Transfer/Reduce

Courts commonly interpret contractual documents on a contra proferentem basis—that is, against the author of the documents—should there be errors, ambiguities, or inconsistencies in the documents. That is, the risk associated with errors in contract documents is generally with the owner/principal.

One way of transferring some of this risk to the contractor is to eliminate some of the contract documents. For example, by not including a bill of quantities, contractors may have to develop their own bill of quantities. The risk associated with errors in this document has been transferred to the contractor. Claims through errors in the bill of quantities no longer exist.

The control used is shifting the onus for document preparation to the contractor. It reduces the objective function—risk associated with claims involving documents.

Example: Transfer

In electricity retailing, one of the major uncertainties faced relates to the purchase of wholesale electricity on the electricity market. Prices that retailers charge to customers are generally fixed, while wholesale market prices can and do fluctuate substantially. The market price level at any time is uncertain, and so retailers face the risk associated with paying far more for electricity than they are able to sell it for. This risk can be transferred through the use of derivative contracts—such as swaps or options.

The control used is the use of swaps and options. It reduces the objective function—risk associated with the electricity purchase price.

Example: Accept

The private sector commonly insures its facilities against damage, theft, etc. This is a transfer of risk. The public sector commonly insures itself. That is, if a public sector facility is damaged, the loss is accepted.

The control used by the private sector is the purchase of insurance. It reduces the objective function—risk associated with damage, theft, etc. The public sector chooses to do nothing. This leaves the objective function unchanged.

Example: Transfer

Much has been published on so-called onerous contracts. These are contracts where the owner/principal writes the contracts in such a way that the contractor bears most of the risk, whether or not the contractor is able to influence these risks. The risk allocation is unfairly biased against the contractor. An example is where the owner does the design, but the contractor carries the risk associated with design errors.

Recommended practice in writing contracts is to have a fair allocation of risk between owner and contractor such that the party that can influence the risk is asked to bear the risk. However, this is easier said than done, given the difficulty of actually working out the risk allocation of contracts in conjunction with all other project issues.

The control used by the owner is the use of onerous contracts. It is intended to reduce the objective function—risk associated with many things.

Example: Reduce

When large bets are placed with a bookmaker, the bookmaker may decide to accept the risk associated with losing a lot of money, in return for possibly keeping the bet money. Or more commonly, the bookmaker will lay off some of the bet money with other bookmakers.

Insurance companies may operate similarly where the matter insured is unusual. Reinsurance gives partial or complete coverage from another insurer for a risk on which a policy has already been issued.

The control used by the bookmaker (and insurance company) is to lay off bets. It reduces the objective function—risk associated with losses.

Example: Insurance

Transfer/reduce: Private citizens carry all sorts of insurance—health, vehicle, home, home contents, travel, etc. Insurance is perhaps the most common way of transferring risk.

Insurance policies however, now commonly, include an "excess" or "deductible"—claims below this level are not considered; for higher-value claims, the amount of this excess or deductible is payable by the insured and the remainder is paid by the insurance company. Additionally, the insured may lose a "no-claim bonus," which might take several years to reinstate to its full value.

With the presence of the "excess/deductible" and the loss of any "no-claim bonus," some of the risk is still being carried by the insured, even though an insurance policy exists.

With additional premium payment, it may be possible to have all the risk carried by the insurance company.

The control used is the purchase of insurance. It reduces the objective function—risk associated with the cost of bad health, accidents, theft, etc.

Accept: Acceptance occurs when a person decides to "self-insure." For example, health insurance covering the gap between a doctor's charged fee and the scheduled fee may not be permissible by legislation.

Doing nothing leaves the objective function unchanged.

Example: Reduce or Transfer

Ways in which a mining company reduces and transfers risks include the use of joint ventures, insurance, hedging, off-take agreements, capital leasing, and contractors.

The controls used are all of these vehicles. They reduce the objective function—risk associated with a business losing money.

Exercise 5.27

How does risk transfer relate to zero-sum game strategies?

Example: Risk Responding

The movie *Fight Club* (20th Century Fox, 1999) gives an interesting example of risk responding.

The scene is that where a recall coordinator is looking at a burnt damaged vehicle, in which the occupants have died. Should a car be recalled because of a known defect? The recall coordinator's job was to "apply the formula."

> Take the number of vehicles in the field (A). Multiply it by the probable rate of failure (B). Then multiply the result by the average out-of-court settlement (C). A times B times C equals X. If X is less than the cost of a recall, we don't do one.

This is the cold organization responding. But given the horror at the personal (car owner) level, the responding of an individual who owns a car would be different to that of the car company, even given the same data.

Example

In conducting business between countries, one risk source relates to currency fluctuations. A typical approach would involve looking at historical records, examining the economic conditions of both countries, and a sensitivity study for fluctuations of ± few percent. Responding would include specifying a preferred currency(ies) of payment, some up-front payment, insurance, and hedging.

For a situation involving gross movements in currencies, this almost certainly would not be identified. Accordingly, control changes would occur.

Exercise 5.28

.Consider an everyday event in your life, for example, crossing a street busy with cars. The risk relates to your getting injured or even possibly killed.

What practices (controls) can you personally adopt to

- Remove the risk?
- Reduce the risk?
- Avoid the risk?
- Transfer the risk?
- Accept the risk?

What practices (controls) do the road authorities/legislators adopt, with respect to pedestrians, to

- Remove the risk?
- Reduce the risk?
- Avoid the risk?
- Transfer the risk?
- Accept the risk?

Exercise 5.29

Transferring risks will almost certainly cost you money, whereas the occurrence of a risk (which represents directly or indirectly a cost) is uncertain. How do you weigh up a certainty versus an uncertainty?

Exercise 5.30

In writing contract documents, there is a view put forward by many that the party, which is allocated the risk, should be able to look after it. Is it possible to write and dissect contract documents to the degree where a clear-cut allocation (transfer/accept) is possible?

Exercise 5.31

At the intersection of two roads, with traffic flowing in both directions on both roads, the volume of traffic and the number of accidents will determine what the road authority does with the intersection. Some options available to the road authority include

- Do nothing
- Install a roundabout
- Install traffic lights
- Separate the traffic by means of an overpass

Interpret these options in terms of the risk responding of remove, reduce, avoid, transfer, and accept (retain, assume).

Exercise 5.32

You inquire about renting a small car (new cost approximately $20,000). The rental rate is $50/day (plus petrol) for unlimited kilometers of travel; if you have an accident, you pay the first $1500 of the accident repair bill. For an extra $13/day, this $1500 reduces to $300, that is, you only pay the first $300 of the accident repair bill.

Which rental rate would you opt for—$50/day or $63/day? Justify your answer. Does your answer change with the length of the rental period?

Exercise 5.33

Having been told that risk management is a special case of synthesis via iterative analysis, develop a formulation for risk management using synthesis terminology.

5.9 Work Study

Work study attempts to create in people a questioning state of mind in the way they view their current and future work practices. This aware state of mind applies to all phases of a project and on a continuing basis in an organization. Work study involves the critical and systematic examination of work with a view to improvement, the elimination of any nonproductive components, and the elimination of waste. Besides the financial incentives, it attempts to create an attitude of mind about the effective use of people, equipment, and materials (Currie, 1959; International Labour Office ILO, 1969).

Work study is made up of *method study* and *work measurement* (Figure 5.11). Method study breaks down work into its components and questions the purpose and need of each component; it involves, among other things, recording information on the work, critically examining the facts and the sequence, and developing alternatives. Work measurement is concerned with the time periods taken to perform work. Much of work study is formalized common sense, yet it is regarded as an essential tool for engineers.

The steps in work study are the same as those in synthesis via iterative analysis, though different terminology is used and the number of steps might be different. Work study is an established technique that has wide applicability. Reengineering is work study reinvented.

5.9.1 Outline

Work study could be said to be as old as work itself. Formalism however has been attempted as far back as the late 1700s by J. R. Perronet, with later contributions by R. Owen, C. Babbage, F. W. Taylor, H. L. Gantt, and F. and W. Gilbreth among others. Of these, perhaps Taylor's "time study" and his "scientific management," Gantt's "Gantt chart," and Gilbreths's "motion study" are best known to most students of engineering. Work study appears to have reached its peak in popularity in the 1940s and 1950s with a steady interest ever since.

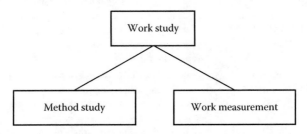

FIGURE 5.11
Method study and work measurement: coordinated procedures.

The end aim in most work study endeavors is to improve productivity. Productivity is the ratio between output and input. It may be described in terms of output (production) per resources used to give that output. The resources may be labor, capital, materials, equipment, or services. Through increased productivity may come higher profit, lower prices, higher wages, etc. The attack on improved productivity comes from several angles including an examination of the basic processes, plant, equipment, materials, buildings, and people.

Work study was widely known for years as "time and motion study," but with the development of the technique and its application to a very wide range of activities, it was felt by many people that the older title was both too narrow and insufficiently descriptive.

The success of work study stems from the fact that it is a systematic way of investigating a problem and developing a solution.

A goal of work study is to improve what already exists. This implies change. The human context of this change is particularly important—how change affects all the people concerned, people at all levels of the team or organization. Change can pose a threat to senior staff, and people generally. There is also the influence of work study on individuals and how they react to being observed.

5.9.2 Method Study

> ... method study involves the breakdown of an operation (or procedure) into its component elements and their subsequent systematic analysis. Thence those elements which cannot withstand the tests of interrogation are eliminated or improved. In applying method study the governing considerations are, on the one hand, economy of operation and, on the other, the maintenance of accepted good practice as laid down by management (e.g. safety and quality standards).
>
> **Currie, 1959**

Method study might also be referred to as *motion study* in some texts.

The general procedure of method study is broken up into the basic steps of

• Select	(the work to be studied)
• Record	(all the relevant facts of the present or proposed method)
• Examine	(those facts critically and in sequence)
• Develop	(the most practical, economic, and effective method, having due regard to all contingent circumstances)
• Install	(that method as standard practice)
• Maintain	(that practice by regular routine checks)

The steps are carried out systematically in order, and all are necessary. Work measurement may need to be carried out in part to assist with obtaining information for these steps.

Some typical areas where method study may bring savings are

- Poor use of materials, labor, or machine capacity, resulting in high scrap and reprocessing costs
- Bad layout or operation planning, resulting in unnecessary movement of materials
- Existence of bottlenecks
- Inconsistencies in quality
- Highly fatiguing work
- Excessive overtime
- Complaints from employees about their work without logical reasons (Currie, 1959)

This provides a hint to the selection process of the work task to be studied.

Recording: Results of recording tend to be displayed visually for easier examination and "before and after" comparisons as charts, diagrams, or models, or a combination of these. A number of recording techniques are used depending on the type of information that is recorded and the detail of the information. More than one technique may be used in any situation. The techniques are not ends in themselves but rather one step in the whole method study.

There are two categories of techniques:

- Charts (for process and time records).
  Within this category, the most generally used are as follows:

Outline process chart	Principal operations and inspections
Flow process chart	Activities of people, material, or equipment
Multiple activity chart	Activities of people and/or machines on a common time scale

- Diagrams and models (for path or movement records).
  Within this category, the most generally used are the following:

Flow diagrams	Paths of movement of people, materials, or equipment
Two- and three-dimensional models	Layout of workplace or plant

Time charts of the (Gantt) bar type are used to represent when activities are being carried out or when resources are required. Time lapse recording and video recording are more recent additions to the armory of method study.

Critical examination: The critical examination is the heart of method study. This takes the form of a systematic examination of the purpose, place, sequence, person, and means involved at every stage of the operation, satisfactory answers being required in turn to each of the following questions:

1	(a) What	(is achieved)?	(b) Why	(is it necessary)?
2	(a) Where	(is it done)?	(b) Why	(there)?
3	(a) When	(is it done)?	(b) Why	(then)?
4	(a) By whom	(is it done)?	(b) Why	(that person)?
5	(a) How	(is it done)?	(b) Why	(that way)?

A satisfactory answer to the query "why?" leads in each case to consideration of alternatives, which might also be acceptable, and finally to a decision having to be made as to which, if any, of these alternatives should apply.

Creative thinking and brainstorming-type techniques may be invoked as a means of generating alternatives.

5.9.3 Work Measurement

Work measurement establishes the time for a person or piece of equipment to undertake a specific task/activity to a given performance level in a given situation. All tasks are broken down into their component or element times. An element is a distinct part of a specified job selected for convenience of observation, measurement, and examination. Times may be established from event recording, element duration recording, or sampling.

A work cycle is the sequence of elements, which are required to perform a job or yield a unit of production. The sequence may sometimes include occasional elements (ILO, 1969).

5.9.4 Reengineering

Reengineering was originally perceived as a radical redesign of an organization's work flows in order to decrease costs and time, and improve effectiveness; work study examines work in all its contexts leading to a systematic investigation of all the factors that affect the efficiency and economy of the work situation being reviewed in order to effect improvement. Reengineering has captured the minds of the trendy; work study grew out of the politically incorrect "time and motion" studies in manufacturing, yet is equally applicable in the service industries and administrative and clerical fields. Many libraries have long since thrown out their 1950s books on work study because of lack of readership; these have been replaced by the tomes of the 1990s management gurus, who appear not to be aware of work study methodology. However, reengineering is work study couched in different terminology.

Exercise 5.34

Reengineering is work study couched in different terminology. As such, the questions are asked: what is new, and why do people embrace reengineering but regard work study as old fashioned and therefore irrelevant?

Exercise 5.35

Having been told that work study is a special case of synthesis via iterative analysis, develop a formulation for work study using synthesis terminology.

5.10 Value Management

> Nowadays people know the price of everything and the value of nothing.
>
> **Oscar Wilde**

Value management (equivalently called value "analysis," value engineering) is popular on projects for a diverse range of situations. Originally conceived as a way of economizing on resources or finding alternatives, it finds application at all phases of a project, from concept through to termination. Central to the approach is a systematic examination of function and alternatives. As such, it can be shown to be but a special case of synthesis via iterative analysis, though different terminology is used and the number of steps might be different.

Its origins trace back to about the 1950s when it was known as *value analysis* or *value engineering*. Then it was a design review or "second look" approach to proposed or existing designs. Its area of application has enlarged over the intervening years such that it now encompasses not only design reviews but also, for example, feasibility studies, an examination of project goals, and conflict situations. The term "value management" is also abused and used incorrectly; for example, the term has been used to describe a public consultation regarding some infrastructure development, including the mediation of conflicting interests.

Central to value management is the examination of function from a whole system viewpoint and the proposing or generating of alternatives. It identifies wastage, duplication, and unnecessary expenditure, and can assist in testing assumptions and needs. The whole system or holistic approach avoids conventional compartmentalized thinking and obtaining locally optimal or suboptimal solutions, at the expense of the desirable globally optimum solution.

Significant cost savings have been reported through the use of value management by a number of writers. From a project viewpoint, the application of value management achieves best results if applied in the early concept and development phases, but it has applicability at all phases and in all tasks.

> **Exercise 5.36**
>
> What is meant by "value"? Could a better term be used to describe what is happening in value management?

5.10.1 Process

Different writers suggest slightly different steps, and terminology for the steps, in the value management process or "value study." A reasonable consensus of steps is information collection, function (primary and secondary) and cost examination, alternatives generation, evaluation of alternatives, and reporting.

The duration of a value study may be anything from a half day to several weeks, depending on the matter being studied. Where prior preparation is necessary, the time span is at the top end of this scale.

An integral part of a value study is a workshop or gathering of interested individuals and stakeholders. The workshop acts as a group problem-solving session with its associated benefits achieved through combining knowledge and accessing the group dynamics. Participants for the workshop are selected such that they represent all disciplines impinging on the matter. Suggested numbers of participants range from say 5 up to about 15, but at the top end, the handling of the workshop and the effectiveness of coming up with new ideas could be expected to decrease.

> **Exercise 5.37**
>
> Having been told that value management is a special case of synthesis via iterative analysis, develop a formulation for value management using synthesis terminology.

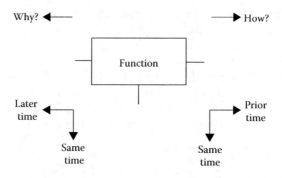

FIGURE 5.12
Diagrammatic description of FAST. (After Dell'Isola, A.J., *Value Engineering in the Construction Industry*, Construction Publishing Corp., New York, 1982.)

Exercise 5.38

Given that value management is but a special case of synthesis via iterative analysis, why is it that modern value management has developed without taking advantage of what the synthesis body of knowledge has to offer? Why has the wheel been reinvented?

Exercise 5.39

In group problem solving, what would be a recommended number of people? On what factors does the choice of number depend?

5.10.2 Distinguishing Features

One feature of value management involves the distinction between primary and secondary functions of a system component or the development of a function hierarchy.

Another is the development of FAST diagrams, FAST here standing for "function analysis system technique" (Figure 5.12). A FAST diagram proceeds from left to right based on the question HOW? and from right to left based on the question WHY?

5.11 Constructability

A constructability or buildability study reviews how something is to be constructed and the construction implications of the design (including the design documents), prior to construction taking place. It identifies and changes anything that may prevent a smooth construction operation, on the assumption that smooth construction should lead to cheaper construction. The study may be carried out after the design is complete or integrated into the design process.

A constructability or buildability study is a blend of work study and value management, applied to construction or building projects. As such, constructability can be shown to be but a special case of synthesis via iterative analysis, though different terminology is used and the number of steps might be different.

Exercise 5.40

Having been told that constructability is a special case of synthesis via iterative analysis, develop a formulation for constructability using synthesis terminology.

6

The Investigation Configuration

6.1 Introduction

This chapter looks at a range of approaches that fit within the investigation configuration, including parameter estimation, state estimation, system identification, characterization, black box, and forecast modeling.

All except state estimation might be grouped within the broader issues of modeling. They are attempting to establish input–output relationships for systems.

The usefulness of modeling is common to all disciplines, not just engineering and the physical sciences. The social sciences, economics, medicine, etc., all have an interest in developing quantitative models for their systems.

Identification approaches range from the complete black box where nothing is known about the system model to the gray box where typically only parameters in the model remain to be estimated (parameter estimation).

Strictly, the version of the model obtained through using input–output pairs is only valid for that range of pairs, although people, as part of human nature, tend to want to generalize the range for which the model is calibrated or derived.

Dealing with black boxes, and in particular the determination of the structure of a model, is very difficult. It is by far the easier to work with gray boxes. What lies inside the black box always remains unsure.

6.1.1 State Estimation

State estimation refers to filtering a system's output in order that some best estimate of a system's state is obtained. The filtering typically removes noise from the output measurements or observations. State estimation assumes a known system model.

Mathematically, for dynamic systems, parameter estimation and state estimation might be treated similarly through the device of calling a parameter a state and developing an equation for the change in the parameter's values over time.

State estimation may be used as a precursor to predicting or forecasting future behavior.

6.1.2 Natural/Artificial Input

Investigation may be approached through the use of natural system input, or the system may be artificially stimulated under deliberate experiments. Typical artificial stimuli include (im)pulses, step functions, ramp functions, and sinusoidal functions.

During investigation, the system ideally should be isolated from inputs/stimuli other than those intended by the observer. In deliberate experiments, this may be possible, while it may be not so in many natural, field, or social systems.

6.1.3 Roots

Much of investigation and optimization has common mathematical roots. Something is being maximized or minimized, as in optimization, in order that the behavior of the model best follows that observed in the system. Hence, what some people refer to as optimization is more correctly called investigation.

6.1.4 On-Line/Off-Line

Investigation can work in two modes:

- Off-line
- On-line or real time

With off-line, system data are firstly collected and then the investigation is carried out. With on-line, the investigation process involves continual updating (of, for example, the system model) as new data come to hand. Typically, with on-line investigation, the investigation process follows a prediction then correction format.

Outline: In the following sections, only some of the more interesting approaches are given. There are many more available in the dedicated investigation literature.

6.2 Black and Gray Boxes

When nothing is known or assumed about the system model, this is referred to as a black box. Black boxes tend to be difficult to work with, and any results might not be practical. They assume nothing of the structure or composition of the system or its potential behavioral characteristics.

As such, it could be argued that the resulting models obtained will tend to be unsophisticated. They also may lack applicability outside the range of input–output data for which they were developed.

Example

Sometimes water catchments are modeled as black boxes, if little is known about the catchment characteristics of soil, vegetation, etc. Rainfall is the input, and runoff is the output. By measuring rainfall and runoff, a model is developed that links rainfall and runoff without worrying about what the physical mechanisms are that connect rainfall and runoff.

Example

Many people will have done a high school experiment in physics to demonstrate Hooke's law. A vertical piece of wire attached at the top has progressively increasing weights added at the bottom end. As additional weights are added, the elongation of the wire is measured. The weights are the inputs, and the elongations are the outputs. If the notions of stress and strain are introduced, this experiment gives the stress–strain (constitutive) relationship for the material in the wire.

Example

Unlike physical illness, mental illness has no seeable causative agent (input), or seeable alteration of a physical organ, or is there necessarily any change in the physiochemistry of the body, which can be used as a basis for diagnosis.

Exercise 6.1

In what situations would you envisage that a black box model would be appropriate?

Generally, it is felt that the better models are obtained by assuming something about the internal makeup of a system and its behavioral characteristics. This is then a gray box. The most common gray box approach is that of parameter estimation, where a model form is assumed and it only remains to estimate unknown parameters in the model from given input–output data.

6.2.1 Parameter Estimation

A number of parameter estimation methods have been put forward in the literature including that of maximum likelihood, least squares, cross-correlation, instrumental variable, stochastic approximation, gradient methods, quasi-linearization, and invariant imbedding.

Many of these require a strong mathematical background and would be unappealing to many people. Least squares is one method that does not require a strong mathematical background, is appealing in its simplicity and ease of comprehension, and hence is possibly the best known and most used method.

Many people will be familiar with least squares style thinking through their high school physics experiments on Hooke's law, where a vertical wire is loaded with weights, and as more weights are added, the extension of the wire is measured. Students are then asked to plot load on a horizontal axis, wire extension on a vertical axis, and asked to draw a line of best fit through the resulting data points.

Anywhere where lines or curves are fitted to laboratory or field data is least squares style thinking. Spreadsheets are equipped to do least squares best fits.

There is also the possible pragmatic approach of guessing parameter values and then checking how good the guesses are by comparing model output with the data. The guesses might then be adjusted iteratively until a good fit is obtained (as is done in synthesis via iterative analysis). In principle, all parameter estimation is trying to do is find the values of some model parameters such that the model's predicted output matches observed output as closely as possible.

Parameter estimation works by minimizing an error measure such that the behavior of the model best fits the system data. The error measure is commonly an output error, that is, the difference between the system output and model output, for the same given input.

That is, tools for parameter estimation range from very sophisticated mathematics through to trial-and-error.

Before embarking on what is usually termed parameter estimation, there are three prior synthesis issues, which generally go unmentioned by writers. These involve

- The choice of the best experiment to conduct in order to get the necessary data
- The choice of the best underlying model form or structure, defined to within certain unknown parameters
- The choice of the best function (error measure) to extremize in order to establish the best parameters

In most cases, writers assume that decisions associated with these three prior issues have already been made, and parameter estimation is then presented as one of minimizing an error measure to give optimal parameters.

Exercise 6.2

Parameter estimation implies that you firstly have to come up with the form or structure of the model. How do you establish which is the preferred form or structure to adopt? Do you use one of a number of standard forms, or do you go back to fundamentals and develop the form from first principles?

6.3 System Response/Output

A system's response or output is a consequence of its input and the system's input–behavior relationship (that is, model).

From experience, it is known that for certain standard inputs, if standard responses are obtained, then the system can be characterized as one belonging to a standard group. This approach is commonly adopted when black box assumptions are used in an attempt to categorize the system model.

6.3.1 Linearity

Models of systems are commonly described as

- Linear
- Nonlinear

Wherever the underlying assumptions are not too extravagant, linear models tend to be preferred because of the ease with which the mathematics can be manipulated. Provided the nonlinear behavior is not too extreme, linear assumptions or piecewise linear assumptions may be made.

Linearity implies two properties:

- Homogeneity
- Additivity

Homogeneity implies that when an input is changed by a factor, the output is also changed by the same factor. That is,

$$\text{if} \quad u(t) \text{ leads to } z(t)$$

$$\text{then} \quad \alpha u(t) \text{ leads to } \alpha z(t)$$

Additivity implies that, for several inputs, the output is the sum of the outputs for each of the inputs:

$$u_1(t) + u_2(t) \text{ leads to } z_1(t) + z_2(t)$$

Any system that does not have both homogeneity and additivity is nonlinear.

Linear equations are linear mathematical models. Nonlinear equations are nonlinear mathematical models. Linear equations tend to be easier to work with, and hence there is a tendency to use linear models wherever possible even though the system may strictly not be behaving with linear characteristics.

6.3.2 Linear Differential Equation Models

Dynamic systems may commonly be modeled by differential equations.

$$a_n \frac{d^n z}{dt^n} + a_{n-1} \frac{d^{n-1} z}{dt^{n-1}} + \cdots + a_1 \frac{dz}{dt} + a_0 z = u$$

Here,
   n is the order of the differential equation
   a_i are constants; $i = 0, 1, \ldots, n$
   t is time
   u(t) is the input
   z(t) is the behavior

For such equations, there is a catalogue of standard inputs and outputs. To use this information, the behavior of a system is observed when subjected to an input, and the differential equation that gives a similar response is used as a model for that system.

Standard inputs include

- Pulse
- Jump
- Sine
- Noise

A pulse is a sudden input of very short duration. A jump is an input applied at a constant level for a finite time.

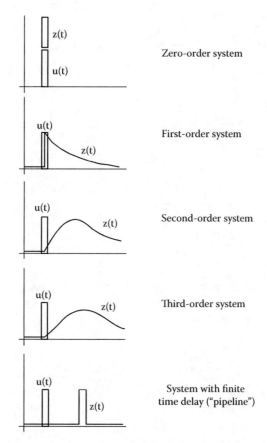

FIGURE 6.1
Some standard system responses/outputs to pulse inputs (impulse responses). (From Kramer, N.J.T.A. and de Smit, J., *Systems Thinking*, Martinus Nijhoff, Leiden, the Netherlands, 1977.)

Typical responses to pulse and jump inputs are given in Figures 6.1 and 6.2, respectively.

Example

Consider a sudden rain shower. This might be regarded as an impulse on the catchment (system). Runoff (response) from the catchment may look something like a lognormal distribution in shape. It might be concluded from this that a suitable model for the catchment is a third-order differential equation.

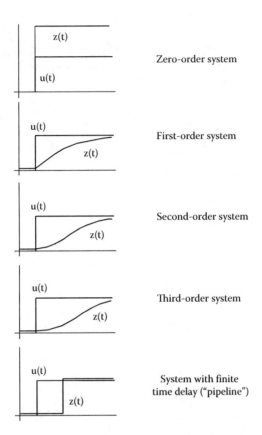

FIGURE 6.2
Some standard system responses to jump inputs. (From Kramer, N.J.T.A. and de Smit, J., *Systems Thinking*, Martinus Nijhoff, Leiden, the Netherlands, 1977.)

Exercise 6.3

Philosophically, what does it mean if some things in nature are said to behave in a linear fashion and some things are said to behave in a nonlinear fashion?

Or is our modeling process wrong? That is, if we developed better representations of systems, would they all be linear models or nonlinear models?

6.4 Least Squares Approach

The least squares approach might be more familiar to some people through the technique of linear regression. Linear regression is an application of least squares theory, which is said to have its origins with Karl Gauss and the prediction of planetary orbits.

It is popular because of its wide applicability, its ease of comprehension, and its requirement of little to no statistical knowledge. *Estimates obtained by the least-squares method also have optimal statistical properties: they are consistent, unbiased, and efficient* (Hsia, 1977).

The results in the literature are usually developed for linear models. Where nonlinearity appears, a transformation to a linear or piecewise linear form may be possible.

The following development is presented in the order of static system models firstly and then extended to dynamic system models.

6.4.1 Static System Models

Consider a (linear) model of the form (Figure 6.3),

$$z = a + bu$$

where
   u is the input
   z is the output corresponding to input u
   a, b are parameters (constants) to be estimated

a corresponds to the intercept on the z axis; b corresponds to the slope of the line.

In least squares, the data may be handled in one block (data grouped) or in a sequential fashion. Both forms are considered in the following text.

6.4.2 Data Grouped

Setting up an error measure as the sum of the squares of the deviation of the data points (u_i, z_i) from the straight line $z = a + bu$ (Figure 6.4),

$$\Delta^2 = \sum_{i=1}^{N} (z_i - a - bu_i)^2$$

the minimum of this measure with respect to a and b can be found. This is known as the method of least squares.

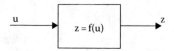

FIGURE 6.3
Input–output representation.

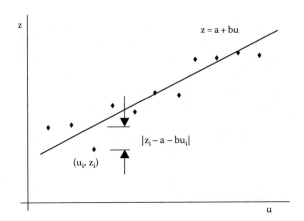

FIGURE 6.4
Least squares/linear regression.

This gives (ignoring some mathematical niceties),

$$a = \bar{z} - b\bar{u}$$

$$b = \frac{\sum (u_i - \bar{u})(z_i - \bar{z})}{\sum (u_i - \bar{u})^2} = \frac{\sum u_i z_i - n\bar{u}\bar{z}}{\sum u_i^2 - n(\bar{u})^2}$$

for the model or regression line $z = a + bu$, where the summations are taken over all the N data points (u_i, z_i), $i = 1, 2, ..., N$, and,

where

$\bar{u}$ is average of all the u_i values
$\bar{z}$ is average of all the z_i values
N is N number of data pairs

Example

Consider the CPI (consumer price index) data in Table 6.1.
The calculations proceed as in Table 6.2.

$$\bar{u} = 55/10 = 5.5$$
$$\bar{z} = 1065.5/10 = 106.6$$
$$b = 32.4/83.0 = 0.39$$
$$a = 106.6 - (0.39)(5.5) = 104.5$$

This gives a model of

$$z = 104.5 + 0.39u$$

which might then be used for prediction or forecasting.
The actual CPI and CPI based on the model are given in Table 6.3 along with the errors.

TABLE 6.1

Example CPI Data

Year	Quarter	CPI Value
1	3	103.3
	4	106.0
2	1	105.8
	2	106.0
	3	106.6
	4	107.6
3	1	107.6
	2	107.3
	3	107.4
	4	107.9

TABLE 6.2

Example Calculations

Period	u_i	z_i	$u_i - \bar{u}$	$(u_i - \bar{u})^2$	$z_i - \bar{z}$	$(u_i - \bar{u})(z_i - \bar{z})$
1	1	103.3	−4.5	20.3	−3.3	14.9
2	2	106.0	−3.5	12.3	−0.6	2.1
3	3	105.8	−2.5	6.3	−0.8	2.0
4	4	106.0	−1.5	2.3	−0.6	0.9
5	5	106.6	−0.5	0.3	0	0
6	6	107.6	0.5	0.3	1.0	0.5
7	7	107.6	1.5	2.3	1.0	1.5
8	8	107.3	2.5	6.3	0.7	1.8
9	9	107.4	3.5	12.3	0.8	2.8
10	10	107.9	4.5	20.3	1.3	5.9
Σ	55	1065.5		83.0		32.4

TABLE 6.3

Example, Comparison of Model and Data

Period	Actual CPI	Modeled CPI	Error
1	103.3	104.9	1.6
2	106.0	105.3	−0.7
3	105.8	105.7	−0.1
4	106.0	106.1	0.1
5	106.6	106.5	−0.1
6	107.6	106.8	−0.8
7	107.6	107.2	−0.4
8	107.3	107.6	0.3
9	107.4	108.0	0.6
10	107.9	108.4	0.5
11		108.8	

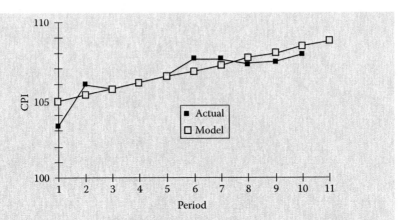

FIGURE 6.5
Example, comparison of model and data.

Figure 6.5 gives plots of actual CPI values and CPI values based on the model.

6.4.2.1 Generalization

The least squares result may be generalized and put into matrix form. The line of best fit,

$$z = a + bu$$

may be written as

$$z = [a \quad b] \begin{bmatrix} 1 \\ u \end{bmatrix} = \boldsymbol{\alpha}^T \mathbf{u}$$

where the vectors

$$\boldsymbol{\alpha} = [a \; b]$$
$$\mathbf{u} = [1 \; u]^T$$

For a more general linear relationship, define a matrix $\mathbf{U}$ and vector $\mathbf{Z}$ that incorporates all the data.

$$\mathbf{Z} = [z_1 \dots z_N]^T$$

$$\mathbf{U} = \begin{bmatrix} 1 & u_1 \\ \vdots & \vdots \\ 1 & u_N \end{bmatrix}$$

Then,

$$\alpha = (\mathbf{U}^T\mathbf{U})^{-1}\mathbf{U}^T\mathbf{Z}$$

which is equivalent but in a different form, to the earlier results.

This requires the inversion of a matrix, as well as matrix multiplication to obtain the values of a and b.

Example

Consider the following data set:

u	1	2	3	4	5	6	7
z	2.9	5.9	8.6	12.1	15.2	18.0	21.2

This is plotted in Figure 6.6.

Fitting a line $z = bu$, then

$$\mathbf{Z} = [2.9\ 5.9\ 8.6\ 12.1\ 15.2\ 18.0\ 21.2]$$

$$\mathbf{U} = [1\ 2\ 3\ 4\ 5\ 6\ 7]^T$$

$$\mathbf{U}^T\mathbf{U} = 140$$

$$\mathbf{U}^T\mathbf{Z} = 421.3$$

$$\alpha = 421.3/140 = 3.009 = b$$

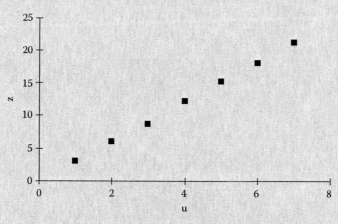

FIGURE 6.6
Example data.

6.4.3 Weighted Least Squares

The least squares estimate earlier is based on weighting every error, Δ, equally.

This can be generalized to allow errors to be weighted differently. Let $\mathbf{W}$ be a weighting matrix. Then

$$\alpha = (\mathbf{U}^T\mathbf{W}\mathbf{U})^{-1}\mathbf{U}^T\mathbf{W}\mathbf{Z}$$

6.4.4 Sequential Least Squares

An alternative approach to estimating the parameters a and b is through a recursive or sequential algorithm.

Consider the case of estimating one parameter, α, according to the line of best fit

$$z = \alpha u$$

Using the earlier least squares result,

$$\alpha = \frac{\sum_{i=1}^{N} u_i z_i}{\sum_{i=1}^{N} u_i^2}$$

For N = 1,

$$\alpha_1 = \frac{u_1 z_1}{z_1^2}$$

For N = 2,

$$\alpha_2 = \frac{u_1 z_1 + u_2 z_2}{u_1^2 + u_2^2}$$

$$= \alpha_1 + \frac{u_2(z_2 - \alpha_1 u_2)}{u_1^2 + u_2^2}$$

Repeating this for N = 3, 4, ...,

$$\alpha_i = \alpha_{i-1} + p_i u_i (z_i - \alpha_{i-1} u_i) \quad i = 1, 2, \ldots$$

where

$$\frac{1}{p_i} = \frac{1}{p_{i-1}} + u_i^2$$

This algorithm requires starting values for α_0 and p_0. p_i reflects the amount of variability in the data. Possible starting values are then $\alpha_0 = 0$ and p_0 large.

Example

Consider a previously used data set:

u	1	2	3	4	5	6	7
z	2.9	5.9	8.6	12.1	15.2	18.0	21.2

Set $\alpha_0 = 0$ and $1/p_0 = 0$. Then,

$$1/p_1 = 0 + 1$$
$$p_1 = 1$$
$$\alpha_1 = 0 + (2.9 - 0) = 2.9$$

Continuing the calculation,

$$1/p_2 = 1 + 4 = 5$$
$$p_2 = 0.2$$
$$\alpha_2 = 2.9 + (0.2)(2)(5.9 - (2.9)(2)) = 2.94$$

Table 6.4 indicates the remaining calculations.

The calculations follow a prediction–correction mechanism where the correction is the right-hand term in the expression for α_i. The decreasing value of p_i indicates increasing accuracy of the estimate for α_i as the calculations progress. The change in α with each new piece of data is shown in Figure 6.7.

Note the calculations process each piece of data in turn and not in one block. New data, as they come to hand, can be readily incorporated.

TABLE 6.4

Example Sequential Calculations

i	p_i	α_i
0	Large	0
1	1	2.9
2	0.2	2.94
3	0.07	2.89
4	0.03	2.96
5	0.02	3.00
6	0.01	3.00
7	0.007	3.01

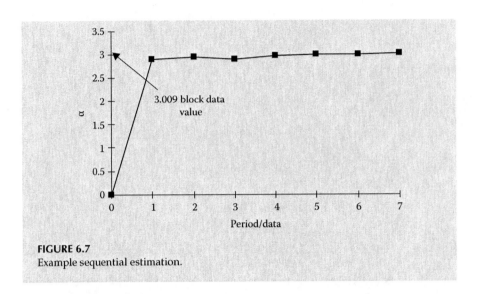

FIGURE 6.7
Example sequential estimation.

6.4.4.1 *Multiple Parameter Case*

For the multiple parameter (constant) case,

$$z = \alpha_1 u_1 + \alpha_2 u_2 + \cdots \alpha_m u_m$$

where m is the number of unknown constants, define

$$\boldsymbol{\alpha}^T = [\alpha_1 \dots \alpha_m]$$

$$\mathbf{u} = [u_1 \dots u_m]^T$$

to give

$$z = \boldsymbol{\alpha}^T \mathbf{u}$$

The equivalent algorithm for the multiparameter case is

$$\boldsymbol{\alpha}_i = \boldsymbol{\alpha}_{i-1} + \mathbf{P}_i \mathbf{u}_i \left(z_i - \mathbf{u}_i^T \boldsymbol{\alpha}_{i-1} \right)$$

$$\mathbf{P}_i^{-1} = \mathbf{P}_{i-1}^{-1} + \mathbf{u}_i \mathbf{u}_i^T$$

where **P** is an m × m matrix. This sequential approach requires the inversion of an m × m matrix.

Suitable initial values might be $\mathbf{P}_0^{-1} = 0$ and $\boldsymbol{\alpha}_0 = 0$, provided $\mathbf{P}_i$ does not end up being singular.

Example

For the earlier example on the CPI data,

$$\alpha^T = [a \quad b]$$
$$u = [1 \quad u]^T$$

Set

$$\alpha_0^T = [0 \quad 0]$$
$$P_0^{-1} = \begin{bmatrix} 1 & 0 \\ 0 & 1 \end{bmatrix}$$

Then

$$P_1^{-1} = \begin{bmatrix} 1 & 0 \\ 0 & 1 \end{bmatrix} + \begin{bmatrix} 1 \\ 1 \end{bmatrix}[1 \quad 1] = \begin{bmatrix} 2 & 1 \\ 1 & 2 \end{bmatrix}$$

$$P_1 = \begin{bmatrix} 2/3 & -1/3 \\ -1/3 & 2/3 \end{bmatrix}$$

$$\alpha_1 = \begin{bmatrix} 0 \\ 0 \end{bmatrix} + \begin{bmatrix} 2/3 & -1/3 \\ -1/3 & 2/3 \end{bmatrix}\begin{bmatrix} 1 \\ 1 \end{bmatrix}\left(103.3 - [1 \quad 1]\begin{bmatrix} 0 \\ 0 \end{bmatrix}\right)$$

$$= \begin{bmatrix} 34.5 \\ 34.5 \end{bmatrix}$$

and so on for P_2, α_2, ...

6.4.5 Nonlinear Regression

Where the type of nonlinearity can be recognized, a nonlinear function can be firstly converted to an equivalent linear function whereupon linear regression may be carried out. For example, the polynomial

$$z = \alpha_1 u + \alpha_2 u^2 + \alpha_3 u^3 + \cdots$$

may be converted to

$$z = \alpha_1 u_1 + \alpha_2 u_2 + \alpha_3 u_3 + \cdots$$

where

$$u_1 = u, \quad u_2 = u^2, \quad u_3 = u^3, \ldots$$

The last expression is now linear in u_i, $i = 1, 2, \ldots$ and the earlier results for the linear case can now be used.

6.4.6 Multiple Linear Regression

Regression models have the potential to incorporate dependencies on multiple variables. For example, future demand may depend not only on time but also on matters such as interest charged on borrowed funds, access to funds, etc. That is, there is possibly some correlation and/or causal relation between the demand and interest rates and fund availability.

For models of the form

$$z = \alpha_1 u_1 + \alpha_2 u_2 + \alpha_3 u_3 + \cdots$$

u_1 might refer to the time period, while u_2, u_3, ... might refer to influencing variables. α_i, $i = 1, 2, \ldots$ are the parameters to be estimated from the regression study. This type of model leads to what is called multiple linear regression. The tools for handling this have been mentioned earlier.

Interrelationships between variables can also be expressed through a collection of regression equations. For example,

$$z_1 = \alpha_1 u_1 + \alpha_2 u_2 + \cdots$$
$$z_2 = \beta_1 u_1 + \beta_2 u_2 + \cdots$$

may be used where there is correlation between z_1 and z_2 but not a causal relationship.

Econometric models adopt such an approach.

6.4.7 Correlation

Correlation is a measure of the degree of linear interrelationship between two variables. Use is made of a sample correlation coefficient

$$r_{u,z} = \frac{s_{u,z}^2}{s_u s_z} = \frac{1}{N} \sum_{i=1}^{N} \left(\frac{u_i - \bar{u}}{s_u} \right) \left(\frac{z_i - \bar{z}}{s_z} \right)$$

where
   $\bar{u}$, $\bar{z}$ are averages of u and z values respectively
   u_i are values of u, $i = 1, 2, \ldots, N$
   z_i are values of z, $i = 1, 2, \ldots, N$
   s_u is the sample standard deviation of u, defined as the square root of the sample variance

$$s_u^2 = \frac{1}{N} \sum_{i=1}^{n} (u_i - \bar{u})^2$$

   ($1/(N-1)$ may be used instead of $1/N$ to avoid bias)
   s_z is the sample standard deviation of z, defined similarly to s_u
   $s_{u,z}^2$ is the sample covariance of u and z
   $r_{u,z}$ is limited in value to $-1 \leq r_{u,z} \leq 1$

For other than a perfectly linear relationship, $|r_{u,z}|$ is less than 1. Where $r_{u,z} = 0$, u and z are uncorrelated (although they may be functionally related nonlinearly).

Example

Consider a previously used data set, reproduced here:

u	1	2	3	4	5	6	7
z	2.9	5.9	8.6	12.1	15.2	18.0	21.2

Here,

$$\bar{u} = 4$$
$$\bar{z} = 12.0$$
$$s_u = 2$$
$$s_z = 6.1$$
$$r_{u,z} \sim 1.0$$

This implies strong correlation. A similar conclusion may have been reached by observing the scattergram (plots of the (u_i, z_i) values) (Figure 6.6).

Exercise 6.4

The following table indicates the cost/m² for various building types. As the rows descend, the cost/m² can be seen to change over time.

Year	Quarter	Factories	Supermarkets	Schools
1	1	432.9	419.6	434.7
	2	441.4	428.0	443.6
	3	450.2	436.8	452.4
	4	461.2	447.6	462.7
2	1	470.8	457.3	472.8
	2	482.9	467.5	485.6
	3	493.7	478.2	496.3
	4	499.1	484.0	501.2
3	1	504.1	488.6	506.0
	2	509.7	494.0	512.0
	3	511.2	497.0	514.5
	4	513.3	500.0	517.1
4	1	513.3	500.0	517.1

Develop a regression equation $z = a + bu$ for the trend (over time) in the costs of either factories, supermarkets, or schools. What do the a and b values refer to?

What is the correlation between the costs of factories, supermarkets, and schools? That is, doing pairwise, what is the correlation between factories and supermarkets, factories and schools, and supermarkets and schools?

Exercise 6.5

This is an exercise on sensitivity. Examine the regression equation obtained for the previous exercise. How sensitive is the cost of the type of building considered to a 10% change in either a or b.

Exercise 6.6

In order to organize an overdraft limit to cover material and labor costs for the coming month, a promoter wishes to predict/forecast sales 1 month in advance. Data on sales are available for the past 12 months, during which time sales have grown by approximately 50%.

The sales data for the past 12 months are

Month	1	2	3	4	5	6	7	8	9	10	11	12
Sales ($ \times 10^3$)	55	50	59	46	64	61	78	85	71	90	67	83

Perform a regression study and fit a line of the form $z = a + bu$ to the data, where z is sales, u is the time period, and a and b are parameters to be estimated.

Compare, for each month, the sales based on the model (predicted/forecast) with the actual sales and calculate the errors.

Plot the actual sales and sales based on the model (vertical axis) versus months (horizontal axis).

6.5 Forecast Modeling

6.5.1 Introduction

Texts on forecasting frequently do not differentiate between the establishment of a model (to be used in forecasting) and using the model for forecasting. The latter belongs within the analysis configuration. The former belongs in the investigation configuration.

Forecasting implies the consideration of future events that might impinge on or influence a project, business, or operation. Forecasting may be used to assist long and short term, strategic and operational planning involving economic matters, inventory levels, workforce levels, provision of services, markets, industry trends, competitor movements, and so on. In production, the intent of forecasting is to help production match future demand. There is a need for forecasting in all areas of planning whether it relates to people, equipment, materials, finances, or other.

As part of overall resource/asset considerations, it is necessary to consider future events that might occur. Is there a market for the completed facility, product, …? Will the return on investment be adequate? Forecasting is also an important part of ongoing operations where costs of readjusting the operations can be avoided, and efficiency of the operation can be improved, if future demand is known. Both long-term and short-term information is required.

Forecasting is carried out by everybody to varying extents. All people use qualitative forecast models, and when no historical data are present, qualitative forecasting is as far as it can be taken. Quantitative forecasting is generally carried out with the aid of mathematical models.

The terms forecasting and prediction are used by many people interchangeably, though some like to make a distinction.

Exercise 6.7

You are asked to provide input to the appraisal and selection stage of the following projects. What role would forecast modeling have for

- The design, production, and marketing of a new vehicle?
- Running for and being elected to public office?
- The conducting of a public seminar?

6.5.1.1 Modeling without Historical Data

Exercise 6.8

How do you develop a forecast model in a situation where you have no historical data?

Common approaches to forecast modeling in the absence of existing data include those based on

- Judgment or intuition
- Polling of experts
- Panel consensus
- Experience/knowledge
- Best guess ("gut" feel)

Where data are available, common approaches to forecast modeling are based on

- Mathematical models
- Graphical eyeball

Exercise 6.9

Assume that you were responsible for forecasting sales of commercial vehicles for engineering purposes. How would you proceed in developing a forecast model?

Exercise 6.10

A company intends to introduce a new small excavator aimed at the domestic construction market. The excavator is essentially a modern-looking version of the basic excavator used in 90% of the domestic market.

What information would be useful in order to model future sales?
What might the model look like?

Example: Design of a Water Recycling Facility

In the design of a water recycling facility, one task is to predict the expected sewage flows of the catchments that will be served by this facility. The modeling involves using the historical flow information and a projection factor to forecast the future sewage flows.

In the case of proposed developments, historical data are not available. Here, the expected flow based on the proposed number of allotments in the area of the development, together with typical average usage per allotment, is used as a basis for the model.

Historical data are a good reference and sanity check, but they do not necessarily reflect future values. The dilemma in any discussion on forecast modeling is "Can the past be used to foretell (the future)?" Whether this can be resolved does not eliminate the need for information on such things as future demand for products, services, etc. Fluctuations in, for example, the stock market may be hard to model, but in engineering and technological areas, it is generally assumed that knowledge of past behavior helps understanding of the future. The demand may need to be known a long time ahead (long time horizon, long-run estimates) or only a short time ahead (short time horizon, short-run estimates). Different time horizons apply to different needs.

Exercise 6.11

What of forecast modeling by analogy? Can you find something similar to what you are trying to forecast and use that as a guide or model?

What of looking at the underlying causes (physical and otherwise) and building up a model from these?

What of "gut feel"? Hunch? Best guess? Appeal to the supernatural? Tarot? Numerology? Some people can "feel it in their bones."

It is important to consider future potential influences. To refine a forecast model, it may be necessary to

- Quantify (based on reasonable assumptions) all foreseeable future potential influences
- Make an allowance (contingency) for other unforeseeable impacts

Example: Managing an Engineering Business Unit

Many people at a company took the days off work either side of public holidays as annual leave. The forecast modeling adjusted down the business unit's financial KPIs (key performance indicators) for these periods.

Other Examples

Other examples of foreseeable potential influences include

- Project schedule—rain days for outdoor projects, procurement delays
- Equipment—breakdowns, maintenance
- People—sick leave, annual leave, turnover, demand

Example: Smallgoods Company

A smallgoods company has been in situations where it has had to develop forecast models for new products, which have no historical data available.

Expert judgment by senior staff, the sales force, and other knowledgeable persons are of some assistance in developing forecast models in situations where historical data are unavailable, such as for seasonal fluctuations.

The company also developed forecast models by looking at the product life cycle analogy of similar products. In order to do this, the company looked at its competitor's experience with a similar product.

The company also incurs considerable expense each year utilizing market research tools such as surveys, mail questionnaires, focus groups, and telephone interviews. The results are used to assist forecast modeling, particularly for products and services that do not have any historical data.

An examination of market trends and the state of the economy assists in forecast modeling. For example, if the central bank increases the official cash rate and interest rates rise, this will lead to a reduction in household income, which will lead to a reduction in overall spending and less demand for products and services.

Exercise 6.12

Consider the upgrade of a dam spillway. The dam design is based on providing for a 1 in 500 year flood event. How can you develop a forecast model for such flood levels without adequate historical data? How can a 1 in 500 year flood event be calculated from only 100 years of data? Could not this lead to massive overspending (or disaster) if the forecast model is even a little bit incorrect?

Exercise 6.13

Some people might argue that there are no situations where historical data are not available to some degree to aid in developing forecast models. The task may be unique, but there is always something that can be used as a frame of reference for the model. Do you agree?

Consider an example of forecast modeling of ticket sales, people movements, etc., for an Olympic Games when there are no equivalent events staged in the host country before. While there is not an event of that size, could previous Olympic Games in another country as well as large sporting events be used to aid forecast modeling? Is it the case that there would always be some data from a similar situation that could be used in modeling in your case?

When forecast modeling for a unique event, are you tempted to skew the model to match a previous similar, but not identical, event? Or do you take only parts of previous data and combine this with current information?

Are many events close enough that the data of the other events can be utilized?

Some Humor according to Scott Adams

Adams (1997) dismisses horoscopes, tea leaves, tarot cards, and crystal balls as *nutty methods* and well-researched sophisticated computer models as a *complete waste of time*. He also argues that assuming current trends will continue into the future is flawed, and uses the example of a pup growing into a dog, which if the trend continued would produce a massive animal. He suggests the following rules *named after myself in order to puff up my importance*.

Adam's Rule of the Unexpected
Something always happens to wreck any good trend.

Adam's Rule of Self-Defeating Prophecies
Whenever humans notice a bad trend, they try to change it. The prediction of doom causes people to do things differently and avoid the doom. Any doom that can be predicted will not happen.

Adam's Rule of Logical Limits
All trends have logical limits.

Extrapolation

Huff (1954) explores the idea of extrapolation.

> Extrapolations are useful, particularly in that form of soothsaying called forecasting trends. But in looking at the figures or the charts made from them, it is necessary to remember one thing constantly: The trend-to-now may be a fact, but the future trend represents no more than an educated guess. Implicit in it is "everything else being equal" and "present trends continuing." And somehow everything else refuses to remain equal, else life would be dull indeed.
>
> For a sample of the nonsense inherent in uncontrolled extrapolation, consider the trend of television. The number of sets in American homes increased around 10,000% from 1947 to 1952. Project this for the next five years and you find that there'll soon be a couple billion of the things, Heaven forbid, or forty sets per family. If you want to be even sillier, begin with a base year that is earlier in the television scheme of things than 1947 and you can just as well "prove" that each family will soon have not forty but forty thousand sets.

A Government research man, Morris Hansen, called Gallup's 1948 election forecasting "the most publicized statistical error in human history." It was a paragon of accuracy, however, compared with some of our most widely used estimates of future population, which have earned a nationwide horselaugh. As late as 1938 a presidential commission loaded with experts doubted that the U.S. population would ever reach 140 million; it was 12 million more than that just twelve years later. There are textbooks published so recently that they are still in college use that predict a peak population of not more than 150 million and figure it will take until about 1980 to reach it. These fearful underestimates came from assuming that a trend would continue without change. A similar assumption a century ago did as badly in the opposite direction because it assumed continuation of the population-increase rate of 1790 to 1860. In his second message to Congress, Abraham Lincoln predicted the U. S. population would reach 251,689,914 in 1930.

Not long after that, in 1874, Mark Twain summed up the nonsense side of extrapolation in "Life on the Mississippi":

In the space of one hundred and seventy-six years the Lower Mississippi has shortened itself two hundred and forty-two miles. That is an average of a trifle over one mile and a third per year. Therefore, any calm person, who is not blind or idiotic, can see that in the Old Oolitic Silurian Period, just a million years ago next November, the Lower Mississippi River was upward of one million three hundred thousand miles long, and stuck out over the Gulf of Mexico like a fishing-rod. And by the same token any person can see that seven hundred and forty-two years from now the Lower Mississippi will be only a mile and three-quarters long, and Cairo and New Orleans will have joined their streets together, and be plodding comfortably along under a single mayor and a mutual board of aldermen. There is something fascinating about science. One gets such wholesale returns of conjecture out of such a trifling investment of fact.

6.5.2 Patterns

When plotted over time, data (or time series—a set of values of a variable given at sequential points in time) may exhibit certain identifiable characteristics. Common characteristics are shown in Figure 6.8; they are also termed trends or patterns. Combinations of these trends are also possible. Identifying trends assists in forecast modeling.

The trends depicted in Figure 6.8a represent "best fits" or central tendencies of the data. A *seasonal index* may be defined as the ratio of the actual data to average data.

Linear trends may be exhibited over parts of the data such that the total data set may be approximated by a piecewise linear form.

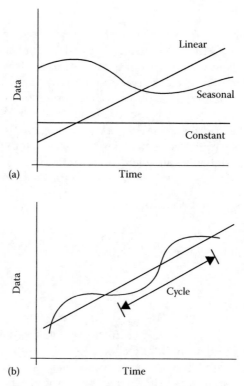

FIGURE 6.8
(a) Standard trends and (b) cycle of a seasonal trend.

Seasonal trends may not necessarily match with the calendar seasons but may have *leads* or *lags* present. For example, preordering of items would result in a lead. A "season" may be 3 months, 1 week, or even a day. For example, activity on a project may have both daily and weekly "seasonal" trends.

Related to seasonal trends is the cycle (Figure 6.8b). Cycles only become evident in large data sets. In effect, cycles are seasonal trends, generally over a longer time scale.

A common method of dealing with seasonal influences is to firstly remove any seasonal pattern from the data, resulting in deseasonalized or seasonally adjusted data. Modeling is then performed. Forecasts based on such models are then seasonalized.

Examples of Patterns

Morning and afternoon urban wastewater flows.
Summer and winter clothing sales.

6.5.2.1 Noise

All data have what is termed *noise* or *disturbances* (Figure 6.9). Noise may be thought of as the departure (dispersion) of the data from a central tendency. With high noise, the data are further away from the central tendency than with low noise. The presence of noise can make the identification of the central tendency difficult and/or lead to errors in forecasts, particularly if done by "eye." An *error* in a forecast is the difference between the value based on the model (forecast) and the actual value when it occurs.

Noise represents variations, sometimes termed random variations, which have no specific assignable cause and no specific pattern. Noise cannot be forecast and is commonly treated separately in forecast models.

Exercise 6.14

Explain what you think are the causes of noise in data.

Example

Consider historical data relating to the flow of wastewater from a series of domestic and industrial catchments.

The tendency of the domestic data was to have a morning and afternoon peak. The main reason for departures from this trend was rainfall. Due to infiltration during and after rainfall, the flow would be higher, sometimes significantly higher.

It was found that flowmeters were sometimes incorrect and in need of recalibration. Meter reset also presented an error. Other errors could arise from telemetry malfunction, data input errors, and human error.

In the case of industrial catchments, trends were more difficult to predict, and it was more difficult to identify the central tendency of the data. The main industrial catchment provided 24 h flow almost 365 days per year. With a central tendency somewhere above zero, it was found that for several hours or days at a time, zero flow would occur when the industry was shut down for maintenance.

Example: Earthmoving

Noise can be caused by foreseeable, but not considered, events. For example, daily earthwork quantities moved in civil projects are heavily affected by weather. Its influence is not usually explicitly incorporated into forecast models.

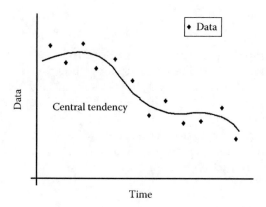

FIGURE 6.9
Central tendency and noise.

Exercise 6.15

Do imperfect assumptions in modeling contribute to noise? Consider the previous examples on wastewater flow and earthmoving— assumptions as to the type (density, hardness, etc.) of material being excavated, or assumptions on the rainfall–runoff relationship.

Example: Measuring the Movement between Rockbolts

One factor that leads to noise in the data is the accuracy of the measuring instrument. In a tunnel, there is a tendency for all rockbolts to move in parallel a certain distance per day. However, when a measurement goes totally against this tendency, it is examined to establish whether it is a real movement or it contains noise.

Example: Abutment Monitoring

One hundred survey monitoring points were established to monitor 70-year-old brick railway abutments. There was a 5 mm accuracy in all three direction coordinates.

Different results occurred depending on the time of day. The noise was caused by the temperature of the survey instrument. Some monitoring points, 100 m inside a tunnel, had a different temperature to that where the instrument was set up. The results depended on the temperature. The morning results and the mid-afternoon results were different. Subsequent examination showed common results for the same time of day.

Exercise 6.16

For manufacturing, output is measured on a daily basis, and there could always be expected to be a variation either on a positive or negative side of the planned output. Whereas nobody is going to question a positive variance, a target may be set to achieve 95% of planned output, otherwise an explanation is required.

Are the following causes of noise in data: machine stoppages due to model change, random quality rejects, and newly recruited operators manning the production line not performing at an optimum level?

6.5.2.2 Stationarity

Other descriptors of the data plotted over time are the terms *stationary* and *nonstationary*. *Stationarity* implies that the trend remains the same over time; another term for this characteristic is *stable*. Conversely, nonstationary behavior is referred to as being *dynamic* or *unstable* (Figure 6.10).

The scale on the vertical axis of such figures can influence a person's view of any pattern or trend present, similar to the way the choice of the horizontal axis scale in histograms can.

6.5.2.3 Dependence

Where the future values of two or more items interrelate, they are said to be *dependent*, and where they do not interrelate, they are said to be *independent*. Where dependency exists, it may be possible to link the models for the two items.

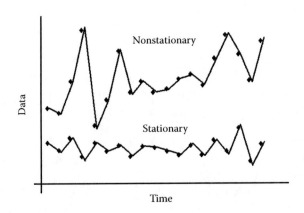

FIGURE 6.10
Stationarity and nonstationarity.

Exercise 6.17

Demand for new two-bedroom and three-bedroom retirement villas over the past 10 years has been as follows:

Year	2-Bed	3-Bed
−10	21	65
−9	36	55
−8	19	50
−7	30	22
−6	22	37
−5	28	56
−4	28	64
−3	23	77
−2	47	76
−1	29	81

How would you characterize these data?

Using your own intuition, develop a demand forecast model.

Using this model, how good would you expect your forecasts to be for this year, next year, and the year after when sales of new villas are expected to start?

Take yourself back in time to 7 years ago. You have 4 years of data (years −10, −9, −8, and −7). Develop a forecast model. Using this, what is your forecast for year −6? Compare your forecast with actual demand in year −6. What is the forecast error?

6.5.3 Forecast Error

An error in a forecast is the difference between the forecast value based on the model and the actual value when it occurs. However, the error might alternatively be reported in terms of a *mean absolute deviation, bias, standard deviation, cumulative sum of errors,* or *sum of errors squared.*

6.5.3.1 Definitions

Define an error, e, as

$$e = \text{Forecast (based on a model)} - \text{Actual}$$

Note that some writers define the error in a negative sense of the earlier, namely, (Actual − Forecast).

Then:

6.5.3.2 *Mean Absolute Deviation*

Mean absolute deviation (MAD) is defined as

$$\frac{\text{Sum of absolute deviations for all periods}}{\text{Total number of periods evaluated}} = \frac{1}{n}\sum_{i=1}^{n}|e_i|$$

where n is the number of periods.

In each period, the forecast (based on a model) and actual values are compared (subtracted), and these are summed over all periods. The absolute value notation accumulates the error and makes no distinction as to the sign of the errors.

6.5.3.3 *Bias*

Bias is defined as

$$\frac{\text{Sum of algebraic errors for all periods}}{\text{Total number of periods evaluated}} = \frac{1}{n}\sum_{i=1}^{n}(e_i)$$

6.5.3.4 *Mean Absolute Percentage Error*

Mean absolute percentage error (MAPE) is defined as

$$\frac{100}{n}\sum_{i=1}^{n}|e_i|/\text{Actual}_i$$

6.5.3.5 *Mean Squared Error*

Mean squared error (MSE) is defined as

$$\frac{1}{n}\sum_{i=1}^{n}\left(e_i^2\right)$$

6.5.3.6 *Comment*

Bias takes into account the sign of the error. A positive bias indicates that the forecast values (based on a model) are generally greater than the actual values. A negative bias indicates the forecast values (based on a model) are generally less than the actual values. Forecasts are said to be biased where they are consistently above or below actual values.

Where the forecast values may sometimes be greater than and sometimes less than the actual values, the mean absolute deviation may end up a much larger number than the bias. The bias measure is more useful where the forecast values consistently overestimate or consistently underestimate the actual values.

The ideal situation is to have both the mean absolute deviation and bias as close to zero as possible. The *magnitude* and/or the *direction* of any error may be important in different circumstances.

Exercise 6.18

For the exercise involving the retirement villas, assume the demand and forecast values (based on a model) for the two-bedroom villas are as follows:

Year	Demand	Forecast (Based on Model)
−5	28	23
−4	28	27
−3	23	26
−2	47	26
−1	29	29

Calculate the mean absolute deviation, bias, mean absolute percentage error, and the mean square error.

Do any of the measures tell you more about the forecast (based on the model) than the other error measures?

6.5.4 Models

There are a number of models available to assist with forecasting, ranging from the completely intuitive through to formal mathematical models. Experience, guesses, hunches, and judgments can assist.

Some of the mathematical models are very involved and are only warranted where accuracy might be obtained and is wanted. The purpose of the forecast will determine which particular model is adopted.

Forecast models might be classified as

- Qualitative (obtained through, for example, Delphi method, market research, historical analogy, surveys, consumer panels, test marketing, and expert opinion/judgment)
- Quantitative (obtained through, for example, moving averages (MAs), exponential smoothing, Box–Jenkins approach, autoprojection, and a regression study)

6.5.4.1 Choice of Model

The qualitative models tend to be suitable for long-range forecasting or forecasting in an area relatively sparse on data. The time series-type models tend to be suitable for short-term forecasting, while causal models might be used for short- and medium-term forecasting. Typically long-range forecasts require less accuracy.

Other factors that influence the choice of model include

- Availability of past data
- Cost (both forecast model development cost, and cost of inaccurate forecasts)
- Underlying demand pattern
- Forecast errors
- Degree of noise

The cost of inaccurate forecasts includes, for example, the cost of over- or under-stocking, over- or understaffing, loss of client goodwill, and so on.

An advantage of using qualitative forecast models is that they have a ready ability to deal with changes. Those involved in the day-to-day operations of a company can anticipate and plan for these changes. This is something that the reliance on historical data and quantitative approaches is unable to accomplish.

A judgmental perspective needs to supplement any forecast model, even the most involved mathematical models.

Care has to be exercised that any forecast model does not permit a forecast overreaction to any sudden increase or decrease in the value of data, particularly recent historical data, which may contain some erratic variation. The forecast model has to reflect any trend no matter how severely the last available data may depart from earlier data.

These conflicting requirements of *stability* and *responsiveness* can be handled to a certain extent by smoothing the data and including several past data points in the development of the forecast model.

As well, experimenting with the available data and calculating forecast errors can assist in establishing the best forecast model.

There are numerous books dealing with forecast modeling and models. Only a few of the models and approaches are described here. A number of off-the-shelf computer packages are available to assist the development of forecast models.

Be aware that two people using the same model form on the same data may arrive at different model specifics and hence different forecasts. Mathematical models involve constants, coefficients, and parameters selected by the engineer. The choice of these determines the specifics of a model and associated forecasts.

Generally, you will never get forecast models that lead to perfect forecasts, with or without past data, but some models will perform better than others.

6.5.5 Expert Judgment

Expert and senior executive judgment is widely used by many organizations.

Judgment forecast models have been used in business for a very long time and find particular use in times of change.

In times of significant change, past data will not generally assist in the development of forecast models, yet forecast modeling is still undertaken through qualitative approaches such as expert judgment.

6.5.6 Delphi Approach

The *Delphi* or *group* approach relies on a group reaching a consensus. Members of the group are chosen with diverse backgrounds and linked through the efforts of a coordinator. Personality clashes between team members are avoided by eliminating direct contact between members.

The approach starts with each team member doing his/her own anonymous modeling, based on posed and factual information. The coordinator edits the contributions from the team members, and this forms the basis of further posed information to the team members, who adjust their models and contribute again. The whole process is repeated until the coordinator is satisfied.

> **Exercise 6.19**
>
> What role does intuition have in forecast model development? Is there a place for a mixed intuition–mathematical model approach?

6.5.7 Market Research

Market research involves extracting information from prospective consumers in a target region or outlet. Tools used in market research include mail questionnaires, focus groups, telephone interviews, and field interviews. The results obtained from market research are used to model the market reaction to the introduction, for example, of some new product or promotion.

6.5.8 Mathematical Models

Forecast models outlined here share the use of this notation:

F_i is the forecast (output) for period i, i = 1, 2, …

D_i is the demand (input) for period i, i = 1, …, n

n, m are number of periods

By partitioning available data, and only using part of the data for forecast model development (parameter estimating—*calibrating, fitting*), the remainder of the data (sometimes termed the *forecasting sample*) may be used to check the suitability of the model.

6.5.8.1 "Naive" Model

A simple or naive forecast model looks like

$$F_{i+1} = D_i$$

6.5.8.2 Moving Average Models

6.5.8.2.1 Simple Average

The simple average (SA) forecast model is based on,

$$\frac{\text{Sum of demands for all past periods}}{\text{Number of demand periods}}$$

In symbols,

$$F_{n+1} = \frac{D_1 + D_2 + \cdots + D_n}{n} = \frac{1}{n}D_1 + \frac{1}{n}D_2 + \cdots + \frac{1}{n}D_n$$

n may increase as new data come to hand.

6.5.8.2.2 Moving Average

The MA forecast model is based on,

$$\frac{\text{Sum of demands for last m periods}}{\text{Number of periods used in the moving average}}$$

In symbols,

$$F_{n+1} = \frac{1}{m}\sum_{i=n-m+1}^{n} D_i = \frac{1}{m}D_{n-m+1} + \cdots + \frac{1}{m}D_n$$

m is kept constant, for example, m = 2 or 3 or …

The MA idea carries forward a window through the data. The size of the window is determined by the choice of m. The earlier "naive" approach corresponds to m = 1.

6.5.8.2.3 Weighted Moving Average

The weighted MA model uses each period's demand multiplied by a weight and summed over all periods in the MA,

$$F_{n+1} = \sum_{i=n-m+1}^{n} C_i D_i$$

where C_i is the weighting for period i, $i = 1, ..., n$; $0 \le C_i \le 1.0$

$$\sum_{i=n-m+1}^{n} C_i = 1.0$$

The weighting coefficients C_i are chosen to reflect the engineer's preference for eliminating, downplaying, or highlighting parts of the data. The performance of the model is dependent on the choice of C_i and m.

Other MA models are available.

6.5.8.3 Exponential Smoothing Models

6.5.8.3.1 First Order

The first-order exponential smoothing model is

Forecast of next period's demand

$$= \alpha[\text{Most recent demand}] + (1-\alpha)[\text{Most recent forecast}]$$

$$F_{i+1} = \alpha D_i + (1-\alpha)F_i \quad 0 \le \alpha \le 1.0$$

where α is a weighting coefficient.

The term "exponential" comes about because if the model is expanded over several periods, the coefficients of the demand terms are exponentially distributed. The most recent demand is usually given the largest weight with older data weighted less and less.

The choice of α influences the outcome. Larger values of α (for example, 0.7, 0.8, or 0.9) weight recent demands most heavily; lower values of α (for example, 0.1, 0.2, or 0.3) weight recent demand less, do not overreact to sudden changes, and lead to insensitivity to current data.

The previous first-order exponential smoothing model may be written in a prediction–correction format, namely,

$$F_{i+1} = F_i + \alpha(D_i - F_i)$$

where the second term represents a correction on the predicted or forecast value (first term) being the same for the next period as it was for the present period. The term in brackets is the negative of the forecast error e, as previously defined, and so

$$F_{i+1} = F_i - \alpha e$$

The exponential smoothing model tends to be better (in terms of forecasts) than the MA model.

6.5.8.3.2 Adaptive Exponential Smoothing   Adaptive exponential smoothing allows the value of α to vary according to the underlying demand pattern and any bias observed. The best value of α is chosen at each time period.

One adaptive model adjusts the value α depending on the forecast error or some function of the forecast error. For example, a tracking signal (TS) might be defined as

$$TS = \frac{Bias}{MAD} \quad -1 \le TS \le +1$$

The value of α remains unchanged as long as the TS lies within certain bounds, for example, within say ±0.75. Should the TS go outside these bounds, the value of α is adjusted up or down in order for the forecast to catch up with the actual.

TS can change sign because bias is used in calculating TS. This does not necessarily mean that there is any correlation between the sign of TS and the direction in which α should change.

MAD is used as a normalizing effect here, such that TS lies between +1 and −1. It is actually the bias that is telling whether the forecast is one side or other of the demand. Bias has a sign (positive or negative).

Other exponential smoothing models are available.

Exercise 6.20

Use the SA model to forecast the demand for the retirement villas whose historical demand data were given earlier. Forecast demand this year, next year, and the year after. How good are your forecast estimates?

Take yourself back in time to 7 years ago. You have 4 years of data (years −10, −9, −8, and −7). What is your forecast for year −6? Compare your forecast with actual demand in year −6. What is the forecast error? Compare your answer with the previous similar exercise.

Repeat for the MA model.

Repeat for the weighted MA model.

Repeat for the first-order exponential smoothing model.

Repeat for the adaptive exponential smoothing model. Start from year −10 with an α value of 0.3. Calculate the TS for each period. Should α go outside the bounds ± 0.75, use a new value of α.

6.6 Dynamic Systems

Parameter estimation for dynamic systems is one of establishing best fit parameters in a model from records of the system operation.

6.6.1 Continuous Time Case

A noise corrupted output, $z(t)$, is observed

$$z(t) = h\big[x(t),\ v(t),\ t\big]$$

where (vectors implied),
   h is a nonlinear function
   $x(t)$ is the state
   $v(t)$ is the observation noise
   t is time

The system model is assumed of the form

$$\frac{dx(t)}{dt} = f\big[x(t), u(t), w(t), p(t), t\big]$$

where (vectors implied)
   f is a nonlinear function
   $u(t)$ is the input
   $p(t)$ are unknown system parameters
   $w(t)$ is the system disturbance

Generally, the order of the model is assumed known.
   Schematically Figure 6.11 applies.

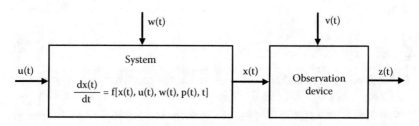

FIGURE 6.11
Schematic of observation and system model.

6.6.2 Discrete Time Case

A noise corrupted output, z(k), is observed

$$z(k) = H\big[x(k),\ v(k),\ k\big]$$

at time instants k = 0, 1, 2, …

The equivalent discrete system model is

$$x(k+1) = F\big[x(k),\ u(k),\ w(k),\ p(k),\ k\big]$$

where H and F are nonlinear functions of the arguments shown.

6.6.3 Kalman Filter

Of all the available approaches to parameter estimation, perhaps the neatest is that given using the Kalman Filter algorithm.

It can be derived from generalizing the sequential or recursive least squares approach given above to dynamic systems, though more complicated mathematical derivations also appear in the technical literature (Gelb, 1974).

Kalman filtering was originally devised for state estimation, that is for determining a best estimate of a system's state from noise-corrupted observations. Parameter estimation can be incorporated within state estimation by regarding the unknown parameters as states. For example, if a parameter, p, is a constant, its state equation can be written as

$$\frac{dp}{dt} = 0$$

or

$$p(k+1) = p(k)$$

The number of state variables (or size of the state vector) is correspondingly greater.

Exercise 6.21

The Kalman Filter and like approaches offer nice treatments of parameter estimation for technical systems. How usable do you believe are the approaches for identification generally, or is general identification too difficult for anything but single-level technical systems?

Examples

Examples from structural dynamics, material creep, material characterization, and constitutive relationships can be found in the following papers by the author: The state estimation problem in experimental structural mechanics (Carmichael, 1979a); Optimal filtering of concrete creep data (Carmichael, 1979b); Identification of cyclic material constitutive relationships (Carmichael, 1980); Adaptive filtering in structural dynamics (Carmichael, 1982).

7

Systematic General Problem Solving

7.1 Introduction

Exercise 7.1

Reflection: If you have never solved a particular problem before and it is of a type that you have not seen before, what processes are involved in coming to a solution? Do you have a specific method or is it approached on an ad hoc basis?

Do you do any of the following? If so, how?

- Pull the situation apart into its components.
- Assemble the situation components into a whole.
- Look at the situation as part of a bigger picture.
- Look at the interaction of the situation with other influences.
- Isolate the situation from other influences.
- Look at causal relationships.
- Try experimentation.

How much is guesswork, luck, and serendipity a part of problem solving? What part does experience, judgment, intuition, intellectual ability, skill, and insight play? What part do existing interpretations of "problem-solving approaches" play?

Do you see a role for having a systematic approach to problem solving to help clarify and define problems, and to help explore solution possibilities?

In problem solving, do you weigh all solutions before arriving at a best choice solution, or do you just go for the first solution that comes to mind?

Do you deal with simple situations without much conscious thought while complicated situations are approached systematically?

Can the problem definition be altered to suit the possible/available solutions? Or is it the other way around? Sometimes possible solutions may look obvious when problems are being defined. However, is the problem definition step trying to know exactly what you are looking at, rather than what you should do in terms of a solution?

7.1.1 What Is a Problem and What Is a Solution?

In broad terms, a *problem* might be described as

> being in a state different to that desired

An alternative state to that existing is sought or wished for. A difference exists between what could or should happen, and what is actually happening.

In broad terms, a *solution* might be described as

> that which transfers the existing state to some other state.

There are degrees of goodness of solutions. Solutions are ranked according to defined objectives. Solutions equate to the selection of controls/decisions/actions that transfer the state from existing to other.

The earlier definitions of a problem and solution are satisfactory for introductory or lay purposes but need tightening up for engineering purposes. More correctly, the state should be expressed as a variable that can take different values. Then a *problem* exists when

> the current values of the state variables are different to that desired

and the *solution*

> changes the values of the state variables

The state variables remain the same from problem definition till after a solution is implemented. The only thing that changes is the values taken by these state variables.

These views on a problem and a solution are quite different to the majority of the literature and peoples' beliefs. The meaning of the term "state" is central to understanding.

Dictionary definitions and lay usages of the term "problem" are rejected here as being unsuitable for developing a systematic framework for problem solving. You will also need to reject such definitions and usages. Typically, dictionaries talk of problems as being "something difficult, doubtful or hard to understand, there being degrees of severity of problems, and a problem being a matter requiring a solution."

Consistent with this, you will also need to banish from your thoughts the use of the term "problem" as encountered in textbooks and classrooms, meaning a contrived (for learning or entertainment purposes) "exercise," "question," or "puzzle," where a "solution" is sought by the text author or class teacher.

7.1.2 Synthesis via Iterative Analysis

Based on observations, the problem-solving process may be broken down to a number of steps that mirror those put forward in synthesis via iterative analysis (and which align with the systems engineering approach of Hall, 1962), namely,

- Definition
- Objectives and constraints statement
- Alternatives generation
- Analysis and evaluation
- Selection

Problem solving and systems engineering involve equivalent steps because they are doing essentially the same thing in systems terms. Feedback and iterative modification, for clarification and refinement purposes, may occur within any of the steps. Each step may uncover previously unthought of issues. Problem solving becomes a process of trial-and-error optimization.

This iteration is not inherent in problem solving, but occurs because of the analysis-based mode of attack.

It may occur that steps overlap. For example, in developing one step, the problem solver may concurrently have in his/her mind information about another step.

The start of the process depends on the recognition of being in an undesired state (that is, the values taken by the state variables are undesirable). The end of the process depends on being in a different (hopefully better) state (that is, the values of the state variables are changed to something more desirable).

Some examples, extensions of those presented earlier, might help clarify problem solving.

Example

Problem: Bank account balance (state) is low, or a higher balance (state) is desired. Objective—for example, (minimum) time to become rich, (maximum) final state. Constraint—for example, current income. Possible solutions (controls): Invest the money at a higher interest rate; deposit more money; etc. The best solution (control) is the one that minimizes/maximizes the objectives, while observing any constraints present.

Example

Problem: A person is unwell (state), or a better health (state) is desired. Objective—for example, (minimum) time to get well, (maximum) final health, (minimum) fuss/complications. Constraint—for example, the cost of medical treatment. Possible solutions (controls): Take medicine; undertake exercise and a special diet; move to a sunnier climate; etc. The best solution (control) is the one that minimizes/maximizes the objectives, while observing any constraints present.

Example

Problem: Person at location A (state) desires to be at location B (state). Objective—for example, (minimum) cost of transport, (minimum) time of travel, (minimum) difficulty. Constraint—for example, forms of transport available. Possible solutions (controls): Drive vehicle; catch public transport; walk; etc. The best solution (control) is the one that minimizes/maximizes the objectives, while observing any constraints present.

Example

Problem: Person is hungry (state), or desires not to be hungry (state). Objective—for example, (minimum) cost to remove hunger, (minimum) time to remove hunger. Constraint—for example, food availability. Solution (control): Anything that removes the hunger or transfers the person from being hungry to not being hungry. The best solution (control) is the one that minimizes/maximizes the objectives, while observing any constraints present.

7.1.3 Abbreviated Version

An abbreviated (and incomplete) version of problem solving is encapsulated in the sometimes heard:

1. Where are we now?
2. Where do we want to be?
3. What is the best way of getting there?

This is most easily understood in terms of transport from place X to place Y, but is sometimes used in discussing career paths of individuals, or strategic aspirations of a company.

Translated, this means (1) define the existing state (that is, the current values taken by the state variables), (2) define the desired state (that is, the desired values of the state variables), and (3) select the values of the control variables to get from (1) to (2) in some best or optimal fashion. It leaves out some important information, however, particularly how "best" or "optimal" is defined (no objectives or constraints are espoused), and how the "way" or "values of the control variables" might be obtained (no model connecting states and controls is espoused).

Example

A quality inspector found a thermal crack in an item. This was symptomatic of a lack of attention to the quenching during the heat treatment process. From a study, it was found that personnel handling the material during quenching are critical, and this defect was due to the unsatisfactory training of personnel in handling the raw materials. From this investigation, the training material and method of training were assessed and improved. Subsequent monitoring shows that the defect type has been reduced considerably.

Original state—presence of cracking
Action/solution—training
Final state—no cracking present

Example

A supplier was not receiving the latest updated information from a company on time, because all information transfer was via email or fax. To address this, the company introduced the use of proprietary software; whenever there was a revision to a document, the software would prompt the user. The supplier was given limited access to company information, while purchase orders were issued online thus ensuring that the supplier was using the latest information to perform the job/project that was assigned to it, and the supplier now had no excuse to say that it had not received the latest information.

Original state—late receipt of information, slow information transfer, waste, poor relationships.
Action/solution—install computer software.
Final state—up-to-date information, fast information transfer, no waste, good relationships.

Outline: The following sections go through each of the steps of problem solving.

Exercise 7.2

The well-referenced example used to illustrate a win-win outcome in negotiation/mediation is that involving two people wanting the same orange. In terms of states, what is the problem?

On the mediator delving further, however, it is found that one person wants only the rind to make a cake; the other wants only the juice to drink. Both people can be satisfied, yet neither gets the whole orange to themselves, and the orange is not cut into halves. In terms of states, what is your revised problem?

Example

In a construction camp, one of the workers had the tap in his bathroom fall apart during the middle of the night. He stopped the rushing water by pulling the copper pipe from the wall and bending it to stop the flow. He reacted before actually thinking to go out and shut the isolation valve outside his unit, or go and find the maintenance person and ask for assistance.

Similarly, with smoke detectors going off in the middle of the night, a few of the camp residents broke the offending smoke detectors by hitting them with the leg of a chair to stop the alarm, or just pulled the detectors from the ceiling.

Exercise 7.3

How does time available to solve a problem, or time devoted to solving a problem, impact on the solution? Do you calculate how long you should spend trying to solve a problem versus benefits of a better solution?

Example: Manufacturing

Typically, when a manufacturing trouble occurs, such as products failing a functional test, an engineer would assess the severity in terms of the failure rate, the criticality of the failure, and the impact on the shipment schedule. If all the aforementioned are within the defined guidelines, no further action would be taken, and production would carry on as usual. However, the engineer would stop a production line if the failure rate was high (greater than 5%) and/or the failure rate might affect the performance of the product.

Next, the engineer would map out the issue, such as defining the root cause (manufacturing process–induced, design-related, or a component-related issue), corrective and preventive actions that could be taken, and target dates for completion.

The time taken and resources used were very much dependent on the severity, criticality, and the impact of the trouble. And the solution may or may not be optimal, depending on the allocated resources and time.

Exercise 7.4

With problem solving, does the Pareto (or 80–20) principle/rule apply? That is, perhaps when solution time is short, can you focus on the part of the problem that is most worrying—the Pareto part that contributes to 80% of the worry? Or does this not make sense?

> One of the nice things about problems is that a good many of them do not exist except in our imaginations.
>
> **Steve Allen**

7.2 Definition

> In all matters, success depends on preparation. Without preparation there will always be failure.
>
> **Confucius**

> If you don't know where you are going, you will end up somewhere else.
>
> **L. Peter**

> A problem well stated is a problem half solved.
>
> **Charles Kettering, 1876–1958**

> A problem well put is half-solved.
>
> **Dewey**

> The formulation of a problem is often more essential than its solution.
>
> **Albert Einstein**
>
> A good start is already towards half of the success.
>
> **Proverb**
>
> In geometry—A figure/sketch well drawn, is an exercise/problem half resolved.

7.2.1 General

At the beginning of the step, a person/group might have a vague notion of a problem and might be unsure how to proceed. To some people, this step involves an acknowledgment that a problem exists, perhaps admitting this for the first time.

Exercise 7.5

What might be used as indicators of a problem existing?

For groups, a common problem has to be recognized and stated.

The definition step involves establishing what the current values of the state variables are, and usually also what are desired values of the state variables, or range of desired values. The state variables remain the same from problem definition till after a solution is implemented. The only thing that changes is the values taken by these state variables.

Exercise 7.6

What is the current state in each of the following cases:

1. "People are not buying our product."
2. A supervisor is concerned about an upsurge in employee resignations.

There can be a temptation for people to rush through this and the following step, and jump straight into giving a solution, even though they may not know what it is the solution of, or whether better solutions may be available. This is much like the way people rush through or bypass planning, and want to start "laying bricks" as soon as possible. There can be a rush to come up with a solution. Quick decision makers are admired. People who think long and hard about a problem are criticized. "Give me your solutions, not your problems." "Let's move on and solve this problem." Yet, it is the preparatory steps that are commonly the most important in any endeavor. Solutions may start to appear but should be put on hold until the problem is properly defined.

The state variables are chosen carefully because the choice influences the direction later taken in attempting to find a solution.

7.2.2 Proper Characterization of the State

In coming to a proper characterization of the current state, people might carry out various activities such as observations, examinations, surveys, data gathering, talking to people, and research generally. Fault and event trees, cause–effect, and fishbone diagrams can be useful for deconstructing the situation.

Example

In a warehouse, there was trouble with stray cats. However, the cats kept the rats away. Indeed, the cats had helped save engaging a pest control company to address rat infestation.

> If all you have is a hammer, everything looks like a nail.
>
> **Anon.**

In principle, the proper characterization of the state might appear straightforward, but the step can be made difficult by the problem solver lacking expertise, or muddied by people issues such as lack of transparency, hidden agendas, or lack of comfortableness with the situation. The question then becomes: has the current state been appropriately characterized?

Example

The gathering of data, although time consuming, helps understand the problem and its cause.

A company encountered inconsistency in the feeding of cover into its assembly machine, which resulted in a frequent jam at the feeding track. There were a number of factors that could cause this trouble, for example, insufficient air pressure, the gap between the cover and track was either too big or small, or dirty cover. The vendor who washed the cover assured the company that the washing process and washing material used to wash the cover had not changed. The company eventually discovered that the vendor had indeed used a different washing material when there was a shortage of the specified washing material. The hidden agenda of this vendor had prevented the team from being able to define the problem well. After matching the data, the company realized that whenever a substitute material was used, there would be feeding trouble.

Exercise 7.7

In terms of defining a problem, would asking the following questions assist: Who? When? What? Where? Why? How? (WWWWWH). This is much like a work study approach, value engineering/management, and also as used in project planning.

Exercise 7.8

How do you know that you have identified the correct characterization of the state, as opposed to a near or related characterization? For example, when a supervisor sees a clash of personalities, the real issue may relate to poor organization structure, and not necessarily relate to people. Similarly, high manufacturing costs may be due to poor engineering design or poor sales planning (and a cost-reduction drive alone might not help).

For example, there is a Dilbert cartoon where an employee is complaining about too much work. So the boss erects a partition in front of the employee, such that the boss can no longer hear the complaints. Problem solved! Or at least it was solved from the boss's viewpoint, but not from the employee's viewpoint, or even the company's viewpoint. This could be the result of poor problem definition, among other things.

Exercise 7.9

How do hidden agendas affect problem definition?

There is a view of human nature that every person has a hidden agenda. This agenda often governs a person's behavior. Other people are often unaware of its origin, but can often feel its existence. Some people are more prone than others. Sometimes a person's agenda is even hidden from himself/herself because they are in denial about it.

Following are some examples of hidden agendas that people may have

- To make themselves feel better (by putting others down)
- To achieve a position of power or prestige (at the expense of others)

- To have power over other people
- To make others feel bad about themselves

Hidden agendas affect the problem definition activity within an organization, because each person is likely to have a personal agenda that differs from others. If the problem definition exposes the weakness of some group members, they may defend themselves by hiding a symptom even if they know what the real cause is.

By not disclosing the factors or causes that contribute to the problem leaves the problem poorly or wrongly defined.

The conclusion is that where people are involved, do not accept the obvious. Always seek out underlying or other issues. If possible, distinguish between people's needs and wants.

Example

Information such as speeds of different cutting tools is important to cycle time improvement. If the true information about the cutting tools is not correctly reported as a result of hidden agendas from the operator/technician, the supervisor diagnoses the wrong cause of the trouble. This leads to wasted effort in addressing the wrong area such as machining method or the CNC (computer numerical control) programming. The reason for the operator/technician hiding the truth may be because he feels that an improvement in cycle time is not beneficial to him, because the improvement will result in an increase in his workload—the number of item/parts to be machined will increase. He may also fear for the reduction in overtime pay as production efficiency improves. Also, he may have the mentality that the project's success is the company's responsibility and not his.

Example

Quality "control" found a batch of food products to be defective. However, checking the process and procedures found nothing wrong. The supervisor in charge was questioned and held responsible for the incident. After a thorough investigation, the company realized that there was a grudge between the process operator and his supervisor. The supervisor had humiliated this operator the day before, and so the operator had secretly added an unknown ingredient to the mixture to take revenge on the supervisor. If this grudge had not been discovered, the real cause of the trouble may have remained a mystery.

Example

There are times when, even after implementing the believed-correct solution to a problem, the problem will still persist. For example, a company had a productivity rate of 88% running continuously for a period of 6 months. No improvement was observed despite increasing the output of the machine by 2% and reducing the downtime from 12% to 10%. Further investigation found that the real cause was the intentional effort of the operators to delay the start-up and early closedown of the process. The hidden agenda of the operators was fear that the increase in productivity would result in the reduction in overtime and thus lead to lower wages.

A symptom may disguise the real underlying problem and its causes.

Example

A headache is a symptom. The cause of the headache may be drinking too much alcohol.

High rates of scrap and rework are usually symptoms of process trouble. The cause of the high scrap rate may be a defective lot of material, a malfunctioning machine, or an inadequately trained operator.

Example

A company faced a quality complaint from customers. The complaint was about battery leakage in some of the pallets shipped. In order to resolve this complaint, the company formed a team to look into the cause. The team implemented a few changes. Unfortunately, complaints still came in. The supervisor decided to send the batteries for further examination. The examination results revealed that the leakage was from the cathode side as opposed to the initial thought that it was coming from the anode compartment. With this information, the company was able to resolve the trouble permanently. The key reason why the problem was defined wrongly in the first place was because the team lacked the expertise to identify the area of leakage due to the seemingly similar symptoms of positive and negative leakage in a battery.

Exercise 7.10

Issues seem to rarely crop up by themselves. Instead, issues can come thick and fast—minor and major issues mixed with each other. How do you uncouple simultaneously occurring issues?

Connotations and people's natural tendency to artificially constrain their own thinking (unable to "think outside the square") can contribute to the "wrong" problem being solved or the wrong solution.

Exercise 7.11

Comment on the following "why" way of approaching a problem.

Example 1: "My car cannot start."

Q. Why cannot my car start? A. Because the igniter cannot start the engine.

Q. Why cannot the igniter start the engine? A. Because the battery is flat.

Q. Why is the battery flat? A. Because I had forgotten to turn off the headlights last night. (Root cause.)

Therefore, should the problem definition be about headlights, flat battery, or car not starting?

Example 2: Before a car wheel can be balanced, any damage to the wheel has to be fixed. Therefore, should the problem definition be about unbalanced wheels, or wheel damage?

7.3 Objectives and Constraints Statement

7.3.1 Objectives

The objectives follow from a developed value system. What is wanted, desired, or needed, that is valued, comprises the value system. These can be difficult issues.

Where alternative solutions are possible, the objectives establish which is the preferred solution.

The difficulty arises because only very rarely do we find it necessary to make an explicit choice of ultimate goals. Our goals and intentions are formed during childhood, adolescence, and even maturity by instinctive likes and dislikes, by our parents, and by our experience in society. And when an explicit choice of goals is called for, there is no one kind of argument which must be presented to validate our stand. This is distinctly different from the situation in the logical sciences (logic and mathematics) where a proof is necessary and sufficient to validate an inference, or in the factual sciences (physics, biology etc.) where observational data and predictive success suffice to substantiate factual claims.

By the statement that there is no one kind of argument to support a given choice of objectives, we mean that there is no body of theory to guide us in choosing objectives. This negative value is probably the most important fact about decision theory in general and value theory in particular. A consequence of this fact often can be observed in engineering reports, where the casualness with which objectives are reported contrasts sharply with the elaborate mathematical models, calculations and empirical data used to justify the choice of one system over another.

Yet it is much more important to choose the "right" objectives than the "right" system. To choose the wrong objective is to solve the wrong problem; to choose the wrong system is merely to choose an unoptimized system.

Hall, 1962

Objectives might not be chosen in isolation, but rather with foreknowledge of an intended solution. As alternative solutions are thought of, so the objectives may change, but this does not have to occur. Objectives may become refined or more detailed or modified as a solution becomes clearer.

Example

Different value systems lead to the following example objectives:

- Profit (immediate, short term, long term)
- Production (product or service)
- Cost (income, economic feasibility, first cost, annual cost—includes the cost of money, depreciation and taxes); cost-effectiveness
- Quality (nonsubjective measures—zero defects, conformance to specification; subjective measures—human response, psychological issues)
- Performance (overlaps with production and quality; figures of merit, efficiency factors, reliability, stability, response, speed, capacity, errors, …; relate to particular systems)

- Competitive issues (retaining or capturing a market segment, affecting or damaging a competitor, minimizing a competitor's profit)
- Compatibility (with existing systems, phase-in periods)
- Adaptability/flexibility (to changing surroundings, ease of conversion, multiple uses)
- Permanence (obsolescence in all its forms)
- Simplicity/elegance (a subjective measure)
- Safety (including probability of system failure, value of consequent loss including loss of life and limb—itself debatable how it is measured)
- Time (including schedule, target dates, lead times, ...)—often implicit in other objectives; duration
- Practicality; ease of implementation; functionality
- Realism
- Consistency with something else (perhaps existing)
- Aesthetics

In addition, there may be combinations of these objectives, for example, production per unit cost.

Some issues may be difficult to measure. Typically, this relates to subjective and so-called intangible issues.

Objectives may be peculiar to a particular individual or organization, because they derive from a value system. Individuals may not be able to agree on what the objectives should be for any particular problem.

Exercise 7.12

Would you expect objectives to be more readily stated for an organization or an individual?

7.3.2 Setting Objectives

There is no unique path to a good set of objectives. Even if one is given a set of objectives, there is no foolproof way to tell whether the set is good. The lack of a comprehensive approach to setting objectives is no excuse for not facing up to the [issues] of setting them. Neither does this lack justify the arbitrariness, imposition, dogmatism and absence of logical thought so frequently found in work on objectives.

There is, happily, some useful and widely accepted theory and philosophy to guide work on objectives. Rather than guide us to the best objectives, the nature of most available theory is that it helps us to spot objectives that are wrong in some sense, or at least which are worse than some others that might be selected to guide the same action. Thus existing aids are rather weak, but it is safe to say that much better results can be achieved by using them than by approaching the [issues] of setting objectives blindly.

Listed below are just a few suggestions that follow from the present state of knowledge about value system design.

(a) Put the objectives on paper. Get agreement that the words used are neutral and free of bias.

(b) Identify means and ends. If this results in several chains of means and ends, "position" the chains in order to locate objectives on the same hierarchical level, and to identify the different dimensions of the value system.

(c) Test to see that the objectives at one level are consistent with higher level objectives. This is necessary to decide the relative importance of various subsets of objectives.

(d) Test that the subset of objectives at each level is logically consistent. Inconsistent objectives signal the existence of trade-off relations. All of these relations will eventually need careful specification to allow for compromises.

(e) Define the terms of trade for related variables. Sometimes all that is necessary is to find the derivative of one variable with respect to another, and to state the limits of the variables within which the trade is valid.

(f) Make the set of objectives complete. The use of experience in similar problems is one way to satisfy this criterion. However, this operation generally continues to the end of design because it is impossible to foresee all consequences of the physical system and cover them with objectives.

(g) Give each objective the highest possible level of measurement. Recognize that some objectives are not measurable on the highest level of measurement scales. Usually the members of certain subsets of objectives can at least be ranked by importance.

(h) Check the objectives to see if each is physically, economically, and socially feasible. State the limiting factors.

(i) Allow for risks and uncertainties by various available techniques and by selecting an appropriate decision criterion.

(j) As a step in settling value conflicts, isolate logical and factual questions from purely value questions. This frequently calls for the use of experts.

(k) Settle value conflicts. Have all interests represented. Use tentativeness. Avoid dogmatism, dictatorial methods and premature voting.

Hall, 1962

Exercise 7.13

Do individuals know their own value systems well enough to explicitly state their objectives?

7.3.3 Constraints

Constraints or restraints restrict the number of possible solutions.

The presence of constraints may simplify the obtaining of a solution by allowing fewer possible solutions, or it may make the solution more difficult because of the added restraints that have to be observed.

Solutions that satisfy the constraints are termed *admissible* or acceptable solutions.

Example restrictions may relate to cost, weight, volume, appearance, operational/maintenance considerations, function, performance, acceptable risk levels, technology, time, capacity, and tolerances.

Further examples include—the solution must

- Be consistent with established policies and practices
- Be acceptable to others; not pose troubles to customers
- Not go beyond the budget; within budget
- Fit within existing workload
- Achieve desired results by year-end
- Not harm the environment (natural)
- Use existing space, existing people, and no more
- Include certain people
- Easy to implement
- Have long-term impact

For workplace problems, constraints might be classified as

- Internal—that is, deriving from departments, groups, supervisors, unions, ...
- External—that is, deriving from customers, government, ...

Exercise 7.14

Many people confuse objectives and constraints. Why is this?

7.3.4 Boundary Conditions

Boundary and terminal conditions may also be present, and these can be regarded, in some cases, as constraints.

7.3.5 Needs

Needs research establishes what the client, consumer, or stakeholder wants, for example, in terms of a new product or service, changed performance level, changed costing, or alternative functions.

7.4 Alternatives Generation

Although possible solutions may have been in the problem solver's mind from the outset, the problem definition step and stating objectives and constraints step set the foundation for obtaining best solutions. The step of generating alternatives now looks at possible solutions. This is the ideas step.

> There is nothing worse than having an idea (if that's the only one you have).
>
> **Anon.**

This alternatives generation step may be routine for established problems. The alternatives are predefined by custom or accepted theories and practice.

Creativity is useful to have in the development of a list of possible alternative solutions. Inventions follow this path. Creativity is discussed later.

But generally, most people follow something else already existing, perhaps with a small change, and obtain solutions similar to that already existing. Previously developed approaches may be extrapolated or interpolated.

Situation: Two people being chased by a wild carnivore.

Problem (state): Imminent death of one of the two people.

Objective: Not applicable. Any admissible solution is acceptable.

Constraint: Death is not allowed.

Solution (by one of the two people): Put on jogging shoes, not to outrun the carnivore, but to outrun the other person.

7.4.1 Environment

Everything not included in the system is the environment. Knowledge of the environment is necessary in terms of how it interacts with the system.

It could be argued that it is not possible to know everything, a priori, relevant to a problem. In such circumstances, the problem solving proceeds with incomplete information. Research, education, and experience can improve this situation.

One approach to considering the environment lists all possible inputs and outputs to the system.

7.4.2 Assumptions

Some assumptions may be needed, but generally assumptions are to be avoided as much as possible.

Assumptions may be hidden or go unrecognized until an effort is made to identify them, and it may be these unrecognized assumptions that prevent the discovery of a good solution. Assumptions may reflect people's values.

> Comedian Benny Hill would have us believe that ASSUME makes an ASS out of U and ME.

7.4.3 Causes

Knowing the cause of a problem can lead to suggestions for solutions. Fault and event trees, cause–effect, and fishbone diagrams can be useful for deconstructing the situation.

> **Example**
>
> The water quality in a dam was of poor quality. It had a bad smell and so was not drinkable. The first attempt at a solution established a deodorization unit in the water treatment plant. This needed significant funds and time to implement. However, after investigation, it was realized that the poor quality was because of sulfide springs near the dam, and their streams flowed into the dam. Most smell occurred in the dry season. A simple solution involved closing these streams into the dam and changing their routes. This was done at low cost and in a relatively short time.

7.5 Analysis and Evaluation

For each alternative generated, an analysis is carried out, and then an evaluation is made according to the established objectives and constraints. Selection
follows in the final step.

The analysis may be based on established models, but day-to-day problems
use reasoned verbal models. The model links states with the controls
(solutions); the current state is already established, and the control is chosen
in order to get to a different desired state.

The form that the analysis takes reflects the form of the objectives and
constraints. For example, if the objectives and constraints are in money
terms, some cost/financial analysis is undertaken using a cost/financial
model.

The form of analysis is chosen to suit the circumstances and may range
from being very mathematical through to being qualitative. A qualitative
analysis, for example, might attach attributes to each of the alternatives.

7.5.1 Uncertainties

Uncertainties might be handled through probabilities or moments (expected
value and variance), obtained either nonsubjectively (for example, historical
data) or subjectively (for example, an expert's opinion or belief).

Sensitivity analyses also assist in this regard.

Exercise 7.15

Why is it important to consider the analysis separate from the evaluation? Or does not it matter?

Matrices might be used as an evaluation tool (Figure 7.1). The objectives/
constraints and the alternatives represent the two axes.

	Objectives					Constraints		
	1	2	3	4	5	(i)	(ii)	(iii)
Alternative A								
Alternative B								
Alternative C								
Alternative D								
...								

FIGURE 7.1
Evaluation matrix.

Sorting of alternatives may be done in a number of ways:

- Duplication and connections: Alternatives that duplicate others might be deleted. Alternatives with similarities might be combined.
- Rank ordering: Alternatives are ranked, and the lowest ranked ones eliminated.
- Categories: Alternatives are placed into different categories. The categories may focus the selection process.
- Candidates for deletion: Some alternatives will not warrant further consideration and can be deleted.

Voting may be carried out in different ways. This applies to group problem solving. Each alternative might be ranked by each group member and collated across all people. The alternative that achieves the best composite ranking is then selected.

Exercise 7.16

Of all the alternatives generated, most will be rejected in any particular problem-solving situation. Is there any value in taking note of these rejected alternatives for future problem-solving activities, or is it better to start the creative thought processes from anew each time?

Should any review of the problem-solving process account for the rejected alternatives?

Exercise 7.17

What other approaches to evaluation can you suggest?

7.6 Selection

The alternative that gives the best values to the objectives, while satisfying the constraints, is the preferred or optimum alternative.

Where there is only a single objective, the selection is straightforward. For example, if (minimum) cost is the objective, all alternatives are evaluated against cost, and the one that gives least cost is chosen.

This example is also deterministic.

The situation becomes more difficult when multiple objectives and/or uncertainty are introduced. With multiple objectives, some subjectivity, reflecting the decision maker's values, has to be introduced. With uncertainty, the calculations become more involved and incorporating uncertainty becomes restrictive on solutions attainable.

Humorous Engineer Identification Test (Source Unknown)

Engineering is so trendy these days that everybody wants to be one. The word "engineer" is greatly overused. If there's somebody in your life who you think is trying to pass as an engineer, give him/her this test to discern the truth.

You walk into a room and notice that a picture is hanging crookedly. You...

a. Straighten it.
b. Ignore it.
c. Buy a CAD system and spend the next 6 months designing a solar-powered, self-adjusting picture frame while stating aloud your belief that the inventor of the nail was a total moron.

The correct answer is (c), but partial credit can be given to anybody who writes "It depends" in the margin of the test, or simply blames the whole stupid thing on "Marketing."

8

Creativity

8.1 Introduction

Creativity is characterized by originality of thought or inventiveness and shows imagination and progressiveness. It involves mental ability and curiosity to come up with a new idea, product, process, ..., development of a new model or theory. At the bottom end of definitions, creativity involves going beyond current boundaries of technology, knowledge, current practices, social norms, or beliefs. But many people would tighten up this bottom end definition somewhat, to insist that some real genuine originality is involved in extending the boundaries, and not just some minor tinkering (for example, changing the color) or doing something different following an exhaustive compilation of what everyone else has done to date (for example, no one may have whistled the national anthem backward).

It is said that creative people have the ability to transcend traditional ideas, rules, patterns, relationships, or the like.

Common definitions of creativity are typically descriptive of an activity that results in producing or bringing about something partly or wholly new; in vesting an existing object with new properties or characteristics; in imagining new possibilities that were not conceived of before; and in seeing or performing something in a manner different from what was thought possible or normal previously.

Creativity is unfortunately one of those imprecisely defined words. Other words with similar definitional issues are, for example, culture and aesthetics.

A distinction might be made between creativity and *innovation*; creativity is *the bringing of something new into being* while innovation is *the bringing of something new into use*. The distinction has a sort of parallel in the terms research and development. Innovation, of course, like any activity, may involve creativity. In fact, some people define innovation as using creativity to add value, where value may be economic, social, psychological, or aesthetic.

The term "innovation" may be used to refer to the entire process by which an organization generates creative new ideas and converts them into novel, useful, and viable commercial products, services, and business practices, while the term "creativity" is reserved to apply specifically to the generation of novel ideas by individuals, as a necessary step within the innovation process. That is, innovation begins with creative ideas. Creativity by individuals and teams is a starting point for innovation; the first is a necessary but not sufficient condition for the second.

Be aware that the term "innovation" has become very faddish, and it has no single meaning in lay usage. It is a term used to impress people rather than to convey a specific meaning or communicate. And some people use the terms "creative" and "innovative" interchangeably.

There are degrees of creativity shown by individuals ranging from being "high" creativity to "nil" creativity. Degrees of originality appear to parallel degrees of creativity.

There is a view that people are either creative or not creative; creativity is a gift possessed by some and held in awe by others; the creative process is mystical and inexplicable, and unable to be understood by noncreative people.

> Most people are other people. Their thoughts are someone else's opinions, their lives a mimicry, their passions a quotation.
>
> **Oscar Wilde**

Exercise 8.1

Is it that some people are predestined to be creative and the rest to be not so? If not nurtured, will creativity disappear from those possessing it? With individuals, is creativity related to heredity or surroundings?

Exercise 8.2

Do some societies suppress creativity in order to control the masses? Consider the popular music culture of the 1960s; George Orwell's book, 1984, where history is rewritten to match the present, and punishment exists for a "thought crime" (human spirit cannot be contained, and it is in people's nature to think creatively); nondemocratic countries; laws and regulations; the training of soldiers.

An alternative view is that creativity involves the relating of previously unrelated matters, an ability to look at something from a new angle or with a fresh eye, and therefore people can be assisted to develop some creativity.

Creativity depends on an inner motivation.

> It takes courage to be creative. Just as soon as you have a new idea, you are a minority of one.
>
> **E. Paul Torrence**

Exercise 8.3

How do you assess whether one person is more creative than another? At an employment interview, how do you assess whether a person has the necessary creativity for a job, or whether the person is creative and may therefore be unsuitable for a job?

Is personal judgment the only way of distinguishing between creative and noncreative individuals?

Is there such a thing as being partly creative?

How do you stimulate any latent creativity that an individual has?

Can training in the creative process be used to increase a person's creative ability?

A general feeling is that various creativity tests provide guidance on an individual's creativity talents but are silent on whether those talents will be used. The effectiveness of an individual in design, advertising, research, or other organizational activity requiring creativity depends on the utilization of the creativity talent. Accordingly, an organization's attention may be directed at bringing out any creativity potential in individuals.

Outline: The chapter looks at the characteristics of creative people, ways of measuring and further stimulating creativity, and the role creativity plays in the life of an organization.

> The reasonable man adapts himself to the world.
> The unreasonable man persists in trying to adapt the world to himself.
> Therefore all progress depends on the unreasonable man.
>
> **George Bernard Shaw**

8.2 Creative Process

Creativity may come about spontaneously, or it could be the result of an active pursuit.

The creative process may involve

- Logic
- Idea linking
- Problem solving
- Free association

8.2.1 Logic

Logical thinking involves the postulation of a model, theory or hypothesis, testing of this, and the drawing of a conclusion. Iterations in this process may occur as experimentation leads to modifying the model or the development of a new model.

Logic involves reasoning of the form

$$if \ A \ then \ B$$

That is, a conclusion is drawn from a set of premises. Where a specific conclusion follows from a set of general premises, this is referred to as *deduction* or *deductive logic*. For example, an apple will fall from a tree because of the law of gravity. Where a general conclusion is drawn from a number of specific instances, based mainly on experience or experimental evidence, this is referred to as *induction* or *inductive logic*. For example, if a number of people die after eating a plant, then the plant may be reasoned to be poisonous.

Exercise 8.4

Give an example of deduction.
Give an example of induction.

8.2.2 Idea Linking

Idea linking involves connecting or relating different pieces of information. Experience and training might or might not enhance the chance of making links; experience and training enlarge a person's knowledge base but also perpetuate established practices and thinking. Education in its purest form, however, should encourage people to raise questions and search for new solutions to problems. Many discoveries are said to be made by people from outside of the area of applicability of the discovery, because such people are not constrained by established thinking in that area.

Aesop Fable: The Three Tradesmen

The Aesop fable, which goes under the name of the three tradesmen or similar, recounts how a bricklayer, a carpenter, and a tanner would fortify a city against an enemy. The bricklayer suggests using brick, the carpenter suggests using timber, and the tanner suggests using leather.
Application: It is difficult to see beyond one's own nose.

Exercise 8.5

List some inventions or discoveries attributed to people from outside of the area of applicability of the invention or discovery.

8.2.3 Problem Solving

A systematic view of the problem-solving process might be presented in terms of the following steps:

- Definition
- Objectives and constraints statement
- Alternatives generation
- Analysis and evaluation
- Selection

Generating alternative solutions may involve lateral thinking, brainstorming, or other free thinking approaches. The best solution is then selected from those generated according to stated objectives and constraints.

Aesop Fable: The Crow and the Pitcher

The Aesop fable called the crow and the pitcher or a similar name recounts how a crow desires to drink the last remaining water in a pitcher, but is unable to because of the shape of the crow's beak. The crow contemplates breaking the pitcher and overturning it, but is unable to do so. The crow then drops a pebble in the pitcher. The water level rises. Each time the crow drops another pebble in the pitcher, the water level rises still further. Finally, the crow is able to drink.
Application: Necessity is the mother of invention.

Exercise 8.6

Consider a particular example of design or planning with which you are familiar. Identify the process components to solving this problem.

8.2.4 Free Association

Free association, rather than systematic, rational thinking, allows freedom from the restraint of logic. Ideas are allowed to be floated unrestrained. The unconscious mind and irrational thoughts are tapped. Censorship is not applied.

8.3 Measuring Creativity

IQ (intelligence quotient) tests have been developed to measure a person's intelligence, though such tests have been criticized. With a similar view in mind, creativity tests have been developed. All tests necessarily have ordinal-type scales. They can be used for comparison purposes between individuals, but the absolute values of the test scores have no meaning.

Examples of the type of things studied in creativity tests are outlined in the following text. Answers given are compared with the frequency of the same answers in a larger population. Answers that are nonsensical are discounted.

- Inkblots—interpret the shape of inkblots
- Anagrams—make as many anagrams as possible of given words
- Unusual uses—consider an object and describe all its possible uses
- Consequences—consider the possible outcomes should a particular event or change occur
- Book titles—read a passage and suggest appropriate titles for the passage
- Thematic apperception—view pictures and develop stories about them
- Word arrangements—given a collection of words, compose a story using some or all of the words

Such tests assess different types of creativity—visual, verbal, ... People may, for example, be more visually creative than verbally creative. However, a person strongly creative in one area seems to also take some of that creativity over into other areas.

Exercise 8.7

a. Of what do the following shapes remind you?

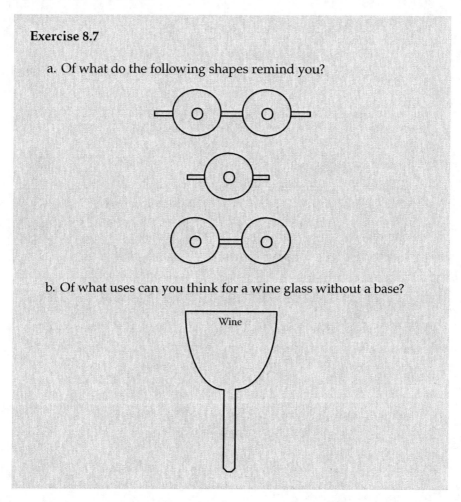

b. Of what uses can you think for a wine glass without a base?

8.3.1 Traits

People found to be creative using these tests quite often have certain other traits: inquisitiveness; dissatisfaction with the status quo; questioning; intelligence; awareness; unconventional in thought and expression; flexibility; independence; purpose; dedication; enriched childhood surroundings; good health; and motivation out of interest rather than reward.

8.4 Types of Creativity

Creativity may manifest itself in various forms including

- Something new
- Combination
- Extension
- A mixture of these

Another form, duplication, is also sometimes included, but its creative nature is questionable.

8.4.1 Something New

"Something new" includes the development of a new model, theory, idea, ... Acceptance of this new thing by peers, who are entrenched in one school of thought or one way of doing things, may be difficult to get; tradition and vested interests may be upset.

> **Exercise 8.8**
>
> List some great engineers and scientists of history who challenged entrenched ways of thinking.
>   Does engineering and science advance by quantum leaps as new ideas emerge or is advancement by incremental amounts slowly changing what already exists?

8.4.2 Combination

Combination implies putting something together from several sources, perhaps unrelated.

> **Exercise 8.9**
>
> Is creativity involved when a supervisor merges the talents of a group of people to produce something that meets the organization's goals?
>   Is there potential for creativity in supervisory tasks?

8.4.3 Extension

Extension implies taking something that exists already and enlarging its usefulness or area of applicability. For example, conceiving and making the first sticky notepads was the result of new thinking. Making them of different sizes, shapes, and colors was an extension based on that original idea. But many people would argue that producing something in different colors or sizes involves negligible creativity. The extension needs to involve some genuine originality before it could be classed creative.

Exercise 8.10

Is creativity involved, say, in bridge building where bridge spans keep on getting bigger? If a black car exists, and someone suggests building the first red car, is this creative? If a car with a manual choke exists, and someone comes up with the idea for an automatic choke, is this creative? If a bicycle exists, and someone suggests installing a (cigarette) ash tray on the handle bars, is this creative? If a bridge that spans 10 m exists, and someone designs a bridge (of the same type) to span 20 m, is this creative?

What degree of enlargement to an existing idea, product, process, ... is required before creativity can be said to exist?

Does taking a theory and developing it into a useful, marketable item involve creativity? That is, in an R&D (research and development) organization, does R involve more/less/same amount of creativity as D?

Exercise 8.11

Consider the automobile industry. Of the following introductions to cars, which involve incremental extensions or genuine creativity?

- The hybrid car
- Aluminum body
- Low drag coefficient—aerodynamic bodywork, flat underbody, low rolling-resistance tires, and extensive use of lightweight materials
- Fuel-efficient car
- Braking systems
- Vision devices that sense
- The use of robots

Is the car today really any different (in substance, not superficially) to the first cars that existed?

The business risks are different between creating something new and extending something that already exists. Business today seems much more comfortable with extension. It is a lot safer, it is incremental, it is building on an already established product or process, and it is far easier to achieve success than starting from nothing.

Incremental improvement in products is a common business goal.

Interestingly, it appears that many of the companies that generate genuinely new ideas, which could change the world, seem to falter in the marketplace.

8.5 Stimulating Creativity

There are two conflicting views as to whether people can be taught to be creative and whether people have latent creativity. However, most people agree that people only use a small portion of their total abilities. Any training, whether under the guise of creativity training or other training, could be argued will help people develop themselves.

8.5.1 Idea Generation

There are a number of methods that can be used to encourage people to generate ideas including

- Taking an idea census
- Systems engineering approaches
- Pure creativity
- Brainstorming and similar

Exercise 8.12

Suggest some other ways alternatives might be generated.

Some approaches to stimulating creativity, either by individuals or groups, are outlined here. Some of the ideas may turn out to be useful, some may not.

> The sublime and the ridiculous are often so nearly related that it is difficult to class them separately. One step above the sublime makes the ridiculous, and one step above the ridiculous makes the sublime again.
>
> **Thomas Paine**

We owe a lot to Thomas Edison. If it wasn't for him, we'd be watching television by candlelight.

Milton Berle

Necessity may be the mother of invention, but frustration is often the father of success.

Anon.

Never reject an alternative because at first it seems foolish or far fetched. A useful technique at this stage is to forget one's critical abilities in the search for ideas. This applies whether one is seeking ideas from others or trying to get ideas from himself. Imagination and evaluation are antithetical to some extent, and these activities can be carried on separately at this stage. Criticism applied too early may inhibit the free flow of ideas.

Sometimes ... the systems engineer must rely on his own imagination and invent a system plan, call upon experts in the relevant field, or abandon the project. It often happens, of course, that the systems engineer does invent, if only because he may be the first technically trained person to encounter the problem. In this case the systems engineer, whose main function is reaching unbiased conclusions, has a new hazard; he must not allow himself to be committed emotionally to his own alternative if other alternatives should appear. He must continue the search for other alternatives to place in competition with his own; he must strive for more than one way of doing a job, if only to dilute his committal to one way.

Hall, 1962

8.5.2 Taking an Idea Census

All known possibilities or alternatives are collected from any available source. With a wide brief, many alternatives might be found; a narrow brief could be expected to severely restrict the range of alternatives.

There are commonly the *do-nothing* alternative and the *status quo* alternative.

In some situations, it may be that there is no existing solution, or no analogy can be found.

8.5.3 Systems Engineering Approaches

Function:

Of all the techniques for aiding the creative approach in planning and design, none is more significant than functional synthesis and functional analysis. Sometimes this technique is oversimplified and called "block diagram design." The technique starts with a statement of boundary conditions, and desired inputs and outputs, and proceeds to a detailed list of functions or operations which must be performed. Then these functions are related, or synthesized, into a system model showing essential logical and time relationships.

Hall, 1962

Delineating subsystems: The function to be performed by the system determines many of the subsystems and subsystem boundaries. Subsystems themselves may be further subdivided in a form of progressive factorization.

[Issues] of interconnecting subsystems generally are reduced when the number and/or number of kinds of inputs and outputs for a subsystem are minimized. This is as true in the design stages, where different design groups have to make the various subsystems compatible, as it is during factory assembly and installation.

Another principle of sectionalization is to minimize the number of interactions between subsystems.

Hall, 1962

Exercise 8.13

Comment on the technique, sometimes referred to as SCAMPERing, as a way of generating solutions:

Substitute: Who else instead? What else instead? Some other material? Another process? Another approach?

Combine: How about a blend or assortment? Combine purposes, units, powers?

Adopt: Does this suggest another idea? What else is like this? Whom could you emulate?

Modify or magnify: What to add: length, height, width, color, topic, shape? New twist?

Put to other uses: New ways to use it? Different techniques or methods?

Eliminate: What or who can you leave out? Is it possible to change an alignment, tighten, streamline, or condense?

Rearrange or reverse: Can you change parts, patterns, layouts, sequence, pace, schedule, people, or places? Can you reverse cause and effect?
  (See Hall, 1962.)

8.5.4 Pure Creativity

> ... criticism inhibits ideation. Sociologists and psychologists have shown that many additional factors inhibit the free flow of ideas. When there is no model of a previously designed system to adapt, no semilogical synthesis techniques, no authority to turn to, the last resort is one's own imagination. In this case, it sometimes helps to know what the inhibiting factors are and to try out a few techniques for overcoming them.
>
> **Hall, 1962**

Exercise 8.14

Do any of the following assist the creativity process:

- Meditation
- Social drugs such as alcohol
- Music (if so, what type and volume)

Exercise 8.15

How much real creativity is exhibited by people, or is it the case that most of the time people copy off someone else's ideas or experiences?

Exercise 8.16

Without taking your pen off the paper, connect all nine dots with four straight lines.

Because this exercise has been around for some time, you may have seen it. If you have seen it, do the exercise with three straight lines without taking your pen off the page, or with one stroke of your pen.

If you had difficulty getting an answer to the puzzle, what were some of the constraints acting on your thinking?

How do you learn to not artificially constrain yourself—"think outside the square"?

Artificial Constraint Puzzles

Consider two puzzles and think about their answers:

1. A pedestrian was wearing all black, even his face was covered in black, walking in a street that had been painted all black, with the buildings on both sides also painted all black. The streetlights were turned off. A car, completely painted black including the headlights, which were turned off, was coming along the street. Somehow the car driver saw the pedestrian and stopped. How could this be?
2. A man and his son were involved in a car accident. The father died, but the son was taken to a hospital for an emergency operation. On getting to the hospital, the doctor said, "I cannot operate on this boy. He is my son." How could this be?

***

Both demonstrate how we artificially constrain our thinking. The first, because of the blackness and the mention of lights, makes us think of nighttime, and we have difficulty making the jump to the answer, which is that all these events occurred in daytime. The second, because of the mention of father and son and boy, makes us think masculine, and we have difficulty making the jump to the answer, which is that the doctor is the boy's mother.

Moral Dilemma

The perfect answer we never thought of.

(Source unknown)

[The way this is phrased leads your thinking. It should be phrased in more neutral terms, such that multiple answers are explored. But it is interesting anyway.]

***

You are driving down the road in your car on a wild, stormy night, when you pass by a bus stop and you see three people waiting for the bus:

1. An old lady who looks as if she is about to die.
2. An old friend who once saved your life.
3. The perfect partner you have been dreaming about.

Which would you choose to offer a ride to, knowing that there could only be two people in your car at one time? Think before you continue reading.

This is a moral/ethical dilemma. You could pick up the old lady, because she is going to die, and thus you should save her first. Or you could take the old friend because he once saved your life, and this would be the perfect opportunity to pay him back. However, you may never be able to find your perfect mate again.

***

Most people think of driving the car themselves, which restricts the remaining transport to one of the other three. However, it is possible to think differently: "I would give the car keys to my old friend and let him take the lady to the hospital. I would stay behind and wait for the bus with the partner of my dreams." Sometimes, we gain more if we are able to give up our stubborn thought limitations, and think outside of the square.

Mindset—False Boundaries and Limitations Created by the Past (Source Unknown)

As my friend was passing the elephants, he suddenly stopped, confused by the fact that these huge creatures were being held by only a rope tied to their legs. It was obvious that the elephants could, at anytime, break away from the ropes, which they were tied to, but for some reason they did not. My friend saw a trainer nearby and asked why these animals just stood there and made no attempt to get away.

"Well", he said, "when they are very young and much smaller we used the same size rope to tie them and at that age it's enough to hold them. As they grow up, they are conditioned to believe they cannot break away. They believe the rope can still hold them, so they never try to break free."

Exercise 8.17

Your task is to make the two sides of the equation equal by changing the position of one line.

$$VII = I$$

What factors may have prevented you from seeing an answer?
What creativity approaches might be useful in reaching an answer?

Exercise 8.18

Produce such a figure "without lifting your pen from the paper":

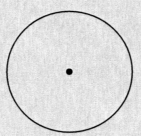

What factors may have prevented you from seeing an answer?
What creativity approaches might be useful in reaching an answer?

8.5.5 Brainstorming and Similar

The most commonly used technique for generating ideas by a group of people is that given (often incorrectly) the name of *brainstorming*. This has a number of versions, for example,

- **Group**: The group is assembled and ideas are contributed to the group and recorded by a central facilitator.
- **Subgroups**: The group is broken into subgroups, which generate ideas as subgroups, and forward to a central facilitator; the facilitator

feeds back the ideas to all subgroups unaltered, or distributes a summary of ideas to all the subgroups, whereby the whole process is repeated.

- **Individuals**: Individuals contribute anonymously their ideas in private to a central facilitator; the facilitator feeds back the ideas to everyone unaltered, or distributes a summary of ideas to everyone, whereby the whole process is repeated.

Central to the approaches is that *no criticism or assessment* is made as the ideas are generated. The assessment comes at a later time; in the "subgroup" and "individual" forms above involving summarizing by the facilitator, some assessment may occur while performing the summarizing, and the facilitator may give a preferred direction to the idea generation.

In general, groups could be expected to generate more and a wider variety of ideas than individuals. However, against this, there is the possibility for *groupthink*, or the domination of the group's thinking by some individuals, whereby the direction that the idea generation goes may get hijacked, and more timid individuals may fail to contribute.

Group sessions can provide a form of communication between levels in an organization, allow ideas of lower rung employees to be heard, and give talented staff the opportunity to be noticed.

> The closest some people come to a brainstorm is a light drizzle.
>
> **Anon.**

8.5.6 Other Techniques

Other lesser known techniques exist including (for example, Hicks and Gullett, 1976; Stoner et al., 1985) attribute listing; input–output technique; grid "analysis"; Gordon technique; Phillips 66 buzz session; catalogue technique; listing technique; focused-object technique; eclectic approaches; and nominal group process.

Exercise 8.19

What value do you see in having regular formal training programs in creativity techniques, or is a more ad hoc approach better?

8.6 Creativity and Organizations

Exercise 8.20

Work is commonly structured to be routine and standardized. Repetition and mass production are sought-after goals. Clearly, creative people would not fit well in such workplaces. The majority of people will be discouraged from being creative.

Does this mean that an organization may have a few creative people with the rest being routine workers implementing the fruits of the creative people?

What if everyone in an organization is creative?

What type of organizations could operate with all creative staff? Are sporting teams such organizations?

Is it possible for people, even doing the most routine of work, to exercise some form of creativity; for example, in improving work output? Or is this taking the creativity notion to its extreme?

Creativity is regarded as an important part of many organizations' existence and functioning. It offers the potential

- For new and better ways of undertaking work
- To gain a competitive edge over rival organizations
- To anticipate change associated with technology

To gain these potential benefits, the particular idiosyncrasies of creative people have to be not only tolerated but also looked after. Companies can maintain a competitive edge by stimulating the creativity of employees.

Exercise 8.21

Is an organization that uses the talents of a creative individual itself creative?

Do organizations have personalities? As such, do creative organizations have the same traits as creative individuals?

Creative organizations are seen to have some common characteristics including creative individuals; open communication; embody suggestion systems; varied personality types; assess ideas based on merit; invest in research; flexible planning; decentralized, enjoyable workplaces; and risk taking.

Organizations can become more productive and competitive through recognizing creativity, encouraging it, and using its end products.

> **Exercise 8.22**
>
> How do you develop creativity in an organization?
>   How do you encourage creativity, while at the same time ensuring some form of discipline and authority in the workplace?

8.6.1 Climate

Creative individuals and creativity training can only be capitalized upon if the organization has the right atmosphere in which new ideas can flourish. Organizations without a favorable culture or climate might stifle any form of creativity.

> **Exercise 8.23**
>
> How would you anticipate an organization's culture or climate to influence an individual's behavior and creativity?

A general feeling is that a person's creativity manifests itself under a combination of circumstances—ability, personality, and climate. The climate needs to be one of support rather than neutrality or even worse hostility.

It is the role of the organization to establish the appropriate climate. This may take several forms including

- The encouragement of new ideas and changes
- Recognizing new ideas
- Maintaining effective communications
- Actively support individual's activities
- Provide the necessary physical infrastructure
- Provide the necessary personal assistance

> **Exercise 8.24**
>
> Creativity implies divergent thinking and a change to the status quo. People generally resist change and can feel threatened by change. How do you encourage creativity in an overall climate that finds comfort in conformity?

Group pressures tend to encourage conformity and discourage people deviating from accepted norms. As such, groups in the workplace can act as a detriment to creativity. It is necessary for groups to adopt, as one of their norms, that change and new ideas can be beneficial, should be considered, and perhaps rewarded.

Exercise 8.25

What issues do you see if employees perceive an unreasonably long time between employee-based idea generation and implementation?

Are the people who generate the ideas necessarily in the best position to comment on the tardiness of their implementation especially when implementation has requirements of resources, human relations, money, marketing, etc.?

Exercise 8.26

List the advantages and disadvantages of concentrating most of an organization's creativity within a single R&D (research and development) department.

Does R&D conflict with ongoing operational work of organizations?

How is an R&D department different to an operational department?

9

General Problem Solving with Groups

9.1 Introduction

Groups are often established for the purpose of solving problems. It might be called a meeting or some more upmarket group name such as quality circles, management teams, task forces, boards, committees, employee groups, ... To advance the problem-solving process, a person, possibly called a facilitator or chairperson, is appointed. The facilitator organizes the problem-solving process, provides structure to something that can become chaotic, and where necessary focuses the group thinking.

Connected issues relate to the physical surroundings, the attitude of individuals and the group, problem-solving culture, agenda, and time issues.

The problem-solving steps are similar for individuals and groups. Groups, however, introduce additional matters relating to how the individuals interact and respond to each other and each other's ideas. Personality clashes and procedural wrangling can interfere with the group process. Meetings, where members become frustrated because of poor chairing, unclear goals, or lack of commitment by others, have been experienced at some time or other by most people; the meetings are plainly nonconstructive.

> A meeting is an event at which the minutes are kept and the hours are lost.
> At the moment of meeting, the parting begins.
> A camel is a horse designed by a committee.

9.1.1 Mediation

A mediator is a neutral person that facilitates negotiations between two or more parties. In effect, through mediation, the conflicting or disputing parties are attempting to settle their differences. Neighborhood, family, environmental (natural environment) and commercial disputes, as well as interpersonal and intergroup conflicts can all be mediated.

Such problem solving differs from more usual problem solving in that there is no single set of objectives and constraints. The mediator attempts to fashion an outcome that the parties can live with, but that may not necessarily satisfy the individuals completely.

9.1.2 Value Management

Value management is an example of group problem solving commonly used in industry. Value management attempts to find solutions, which are cheaper but still perform the required function. Value management workshops involve input from various specializations that impinge on the problem; they are facilitated and run as problem-solving workshops.

A more liberal, but minority, interpretation by some people of value management is that it is a process of reconciling the interests of disparate groups. As such, it is also a problem-solving exercise. Commonly, community consultation workshops, surrounding an impending project, might be called value management workshops in this sense.

Outline: Each of the following sections looks at different aspects of group problem solving.

9.2 Participants

Group problem-solving sessions commonly involve

- A facilitator
- A recorder
- Group members

Some people may take more than one role, depending on the circumstances.

9.2.1 Facilitator

Two options are available in the *selection of a facilitator*:

- The facilitator may be a *neutral* person who solely looks after the process.
- The facilitator may adopt a *dual role* of guiding the process while contributing substance to the process. Commonly, this is the case for meetings in the workplace. Everyone has biases and has difficulty presenting a neutral view, and facilitators are no exception. Where the facilitator is someone with rank, this rank may interfere with obtaining the best solution.

Generally, a consensus view of the preferred type of facilitator is that s/he be a neutral. However, because of workplace size, location, the problem, and the ability of the facilitator, this may not be possible to have. Facilitators have the potential to strongly influence the way that the rest of the group thinks and the direction that the group goes.

Exercise 9.1

How difficult is it for a facilitator to remain neutral when dealing with problems associated with the facilitator's workplace? Can a facilitator wear two hats, sometimes being a neutral facilitator and sometimes contributing to the substance of the process, with switching occurring between these roles?

There is the additional consideration regarding the *facilitator's background*:

- The facilitator may have no subject or technical expertise. Such facilitators will have trouble understanding the discussion, thereby reducing the efficiency of the process. The process may also go off in a fruitless direction because of the facilitator's ignorance.

- The facilitator may be a subject or technical expert. Such facilitators may suffer from having preconceived views and be set in their ways by following industry custom. It may be difficult for them to take a fresh look at problems. Strongly opinionated people may also suffer similarly. They may skip issues that they consider are irrelevant, but which may actually turn out to be relevant.

Generally, a consensus view of the preferred type of facilitator is that s/he be somewhere in between these two extremes. Too many anecdotes exist, in particular, about asking a person with no technical knowledge to facilitate the problem-solving process, and the process has ended up a waste of time and money.

Neutral facilitators may be hired or selected from elsewhere in the organization.

Where a neutral facilitator is not available, an attempt needs to be made by the facilitator to remain neutral. This becomes more important where the facilitator outranks the others in the group.

Exercise 9.2

Is it possible for a facilitator to be lower ranked than someone in the group? Does it work whereby, for the duration of the problem-solving process, the facilitator instructs the boss on correct behavior, while after the process the reverse applies?

Problem Solving for Engineers

Exercise 9.3

Is it possible to carry out group problem solving without a facilitator? Or does someone naturally adopt a facilitator's role even though they may not be called such? Would you expect the solution to be better with a facilitator or without a facilitator? Or are facilitators an expensive luxury?

9.2.2 Recorder

While the process is going on, a recorder takes notes and these act as the *group memory*. The notes may commonly be recorded on a whiteboard, visible computer, or paper mounted around the room. It is a public record; it is part of the process.

The facilitator may adopt the dual role of recorder and facilitator.

Any biases that the recorder may have desirably should not be reflected in any translation or coloring of group views to be more closely aligned with those of the recorder.

Exercise 9.4

Under what circumstances would it be appropriate for the facilitator and recorder to be the same person? Does it depend on the group size or other matters? Do large groups necessarily require a separate facilitator and recorder, or does it depend on the speed at which, for example, ideas are being generated?

The display of the record converts part of problem solving into a visual process. It acts as a historical record of thinking on the problem as new information is added. Being displayed at large, it becomes owned by the group rather than an individual.

The display assists in the organizing, distilling, and synthesizing of information. It provides a datum from which the process continues after breaks or interruptions.

Tricks used by the recorder include

- Headlining (key thoughts or key points) with supporting points
- Abbreviations
- Symbols (mathematical in form)

- Diagrams (for example, people faces to show moods, block diagrams, arrows, ...)
- Highlighting for separation and emphasis

9.2.3 Group Members

For best results, all group members should make contributions. To be avoided is the "shrinking violet" who makes no contribution through timidness or lack of self-confidence, and the "bull" who dominates the process, often talking for the sake of talking and preventing others from making more valuable contributions.

Exercise 9.5

Within a group set up for problem solving, is it possible to have people of different rank and for the group to function properly? Are low-ranked people likely to be hesitant to contribute in front of their bosses? What behavior role should the boss adopt in such situations?

What about an "opposite" behavior where the junior tries to impress the boss, but in so doing detracts from the problem-solving intent of the meeting?

If the option of asking the boss to leave is not available, what needs to be done in order that the problem-solving process proceeds as smoothly as possible?

How might various national cultures deal with differently ranked or classed people in the one problem-solving session?

Example

In implementing a quality standard in one company, numerous meetings were held. These meetings often involved all the departments from the company, with different ranks of staff involved. Lower-ranking staff were always quiet and not participating, especially when the managing director was around. Fortunately, this managing director was good at stimulating conversation, and he ensured that everyone participated by helping them to understand and become enthusiastic about the topic discussed. He created an atmosphere whereby lower-ranking staff felt safe and not threatened during the meetings. He always kept the group goal up front and provided encouragement and motivation for both the steering and working committees of the implementation project.

Differently ranked people within a group contribute different perspectives on a problem. Higher-ranked people may look at problems from a broader perspective, coupled with concern and caution. They also have experience to contribute. Lower-ranked people may look for the detail workface issues in problems.

Conflicts of interest within a group can assist solving problems. Any view, good or bad, can lead to an idea. All ideas generated have the potential to stimulate further ideas. However, when dealing with large groups with conflicting ideas, it may be necessary to have a facilitator who can assess the development of ideas and intervene when the discussion is believed to be tending to unproductive.

9.3 Facilitation

9.3.1 Facilitator's Styles

Two extreme facilitation styles are as follows:

- The facilitator takes a very *prominent proactive* style. When a group has just formed, or where the group members lack maturity, such a style may be necessary in order to promote progress. Such forceful styles can lead the group in a wrong direction.

- The facilitator takes a back seat to the process and only imposes on the process when it is perceived to be going off the rails. The facilitator appears *invisible* to the group. Mature groups may operate with such a facilitator. To many people, it is considered desirable if a group can reach this state.

As a group develops, the facilitation style may shift from being highly visible to invisible. The style might be changed also depending on how long the group has been sitting, how long since the last break, and so on. People have times of the day when they perform better or are more active than other times. The facilitator needs to be aware of this and respond accordingly.

Exercise 9.6

In the role of facilitator, how realistic is it to expect that a person can switch from being visible to being invisible? Will a dominant facilitator always be dominant? Will a weak facilitator always be weak?

9.3.2 Facilitator's Activities

A variety of activities might be expected of a facilitator. These may be broadly categorized as

- Looking after the process
- Handling the people
- Looking after the surroundings

9.3.2.1 Looking after the Process

This might possibly involve the input of suggestions from the group as to a preferred approach.

- *Establish ground rules*; encourage people to accept and listen to others' viewpoints; discourage negative criticism of others' views; support openness, honesty, listening, showing respect, patience with each other, confidentiality, and acceptance of mistakes.
- *Establish the process* to be followed; discourage deviations and the group going on false trails; establish an associated timetable for the process to follow.
- *Identify obstacles* and impasses in the way to a solution and assist the group overcome them; various tricks are available to assist overcoming obstacles such as taking breaks, breaking into smaller groups, introducing humor, changing direction, and introducing new personalities.
- *Identify common threads*; this helps to focus and provide system and convergence to the process rather than leading possibly to damaging divergence; recognize any consensus and reinforce this—consensus may build on itself, that is, some consensus may lead to further consensus; recognize when the group is in agreement and when some issue has the support of everyone.
- *Perturb the process* if nothing significant appears to be happening; various tricks are available to perturb the process such as taking breaks, dividing the group into smaller groups, introducing humor, changing direction, introducing new personalities, introducing new information, play the devil's advocate—present alternative viewpoints, create straw issues that are intended to be shot down, ask questions, redirect questions and statements back to the group, and go from generalities to specifics, or from specifics to generalities.

- *Summarize* the group's endeavors along the way and at the end.
- *Examine, debrief, reflect, and learn* about the problem-solving process; take this information to future problem-solving sessions.

Exercise 9.7

Add other activities expected of a facilitator in looking after the process.

9.3.2.2 Handling the People

- *Encourage participation* through establishing a participative climate; allay fears of some group members that their contribution is not important or as relevant as others' contributions; perhaps discourage the overtalkative ones and encourage the quieter persons to speak.
- *Motivate the group* and reward the group for its contributions.
- *Resolve conflicts* within the group; focus attention on the issues and away from the people; arrange seating to reduce conflict.
- *Be aware of the group's feelings*, sometimes called the group's pulse; dispel fears that a solution may not be reached; dispel possible fears about the futility of the exercise.

Exercise 9.8

Add other activities expected of a facilitator in handling the people.

9.3.2.3 Looking after the Surroundings

One task of the facilitator is to overview the physical surroundings of seating, lighting, ventilation, recording materials, refreshment, and eating facilities.

9.3.2.3.1 Seating

For problem solving, it is recommended that the focus of the group be the whiteboard/flipcharts, such that seating might be something like Figure 9.1.

Seating around a table or across a table focuses the group somewhere in the center of the table, where there is no feedback to the group. Attention shifts

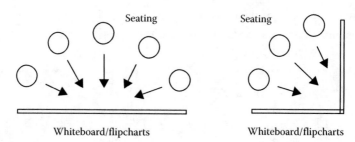

FIGURE 9.1
Suggested seating patterns.

to looking at other people rather than at the problem. Information is transferred between individuals rather than to the group knowledge contained on the record. Personalities and interpersonal relationships start to interfere with the process. The seating in Figure 9.1 shifts the focus from the people to the problem.

9.3.2.3.2 Room

Considerations regarding the room include

- Size (not too big or too small)
- Ventilation, temperature
- Lighting
- Chairs—number and comfort
- Tables—number
- Marking pens and paper
- Refreshments, meals—go out or eat in

9.3.2.4 Agenda

The agenda sets the tone for the meeting. It may be developed by the facilitator in conjunction with other interested parties. The agenda conveys to the group the purpose of the meeting, how the problem will be attacked, and the time available.

Consideration should be given to

- Stating the purpose of the meeting
- Inviting the correct people to take part: no more, no less
- Involving people with the necessary authority to carry forward any solutions developed

- Involving experts, on an as-needs basis
- Distributing the agenda and relevant material beforehand in order that the group is prepared
- Time frame allowed
- Hidden agendas that might affect the process

Exercise 9.9

Add other activities expected of a facilitator in looking after the surroundings.

9.3.3 Facilitator's Skills

Skill requirements of facilitators are numerous and include

- A knowledge of *body language* for use by the facilitator and for interpreting signals from group members
- *Listening skills*; listening to the group's ideas rather than imposing the facilitator's ideas on them
- *Summarizing skills*
- *People-handling skills*

Exercise 9.10

How do you learn facilitator's skills?

9.4 Problem-Solving Steps

The problem-solving steps are the same whether an individual or group is involved, namely,

- Definition
- Objectives and constraints statement
- Alternatives generation

- Analysis and evaluation
- Selection

with possible feedback between steps.

There are some practices, which assist problem solving with a group.

The process starts with introductions and an explanation of the steps and procedures to be followed.

For the definition step, the facilitator might adopt a more structured, more formal approach until the group warms to the occasion.

With a mutually agreed definition, there is the potential for commitment from the group to a solution.

It is important that all members of the group contribute and become involved in the process right from the start. The task of the facilitator is to ensure that this occurs.

People who are present but do not wish to be, or people who see no point in the particular problem-solving exercise, or people who like to be part of a group for its sake only can be asked to leave.

It is important that the correct mix of skills is present in the group and that the people able to commit to outcomes are present.

All people need to be present from the start of the process in order that all people are fully informed of everything that happens and no backtracking occurs.

There can be a desire by some people to shorten the definition step in order that they can "get into it." However, it is argued that by building a strong foundation in this step, the later steps progress more effectively.

If solutions start to emerge, put these on hold until the problem is fully defined or record them separately.

The pace of the definition step should be steady, subject to involving everyone and not becoming too focused too early.

Questions can be the most effective tool used by the facilitator. Preferably, the questions are open, in order that group members can express their own views.

Data and information, on which the group is to work with, need to be accepted by all. Consensus builds on agreement.

In the alternatives generation step, many people have trouble thinking beyond what already exists. It is the facilitator's role to encourage creative thinking, to encourage the group to come up with as many alternatives as possible.

The objective will need clarification, in terms of whose value system is involved—individuals or organizations—and for the evaluation step, if multiple parties are involved, how each will be weighted.

9.5 Groups versus Individuals

The Blind Men and the Elephant (Bowen, 1987)

It was six men of Indostan
To learning much inclined
Who went to see the Elephant
(Though all of them were blind)
That each by observation
Might satisfy his mind.
The First approached the Elephant
And happening to fall
Against his broad and sturdy side
At once began to bawl:
"God bless me! but the Elephant
Is very like a wall!"
The Second, feeling of the tusk,
Cried, "Ho! what have we here
So very round and smooth and sharp
To me 'tis mighty clear
This wonder of an Elephant
Is very like a spear!"
The Third approached the animal
And happening to take
The squirming trunk within his hands,
Thus boldly up and spake:
"I see," quoth he, "the Elephant
Is very like a snake!"

The Fourth reached out an eager hand,
And felt about the knee
"What most this wondrous beast is like,
Is mighty plain," quoth he;
" 'Tis clear enough the Elephant
Is very like a tree!'

The Fifth who chanced to touch the ear
Said: "E'en the blindest man
Can tell what this resembles most;
Deny the fact who can,
This marvel of an Elephant
Is very like a fan!"
The Sixth no sooner had begun
About the beast to grope,
Than, seizing on the swinging tail,
That fell within his scope,
"I see," quoth he, "the Elephant
Is very like a rope!"
And so these men of Indostan
Disputed loud and long,
Each in his own opinion,
Exceeding stiff and strong,
Though each was partly in the right,
And all were in the wrong!

The Moral:

So oft in theologic wars,
The disputants, I ween,
Rail on in utter ignorance,
Of what each other mean,
And prate about an Elephant
Not one of them has seen!

John Godfrey Saxe, 1816–1887

Problem solving requires involvement and commitment. Through this, the group can focus on the path to a solution and gain ownership of the process and the result.

It is the skill of the facilitator that can harness the participation of the group members, particularly people who have not been involved in problem-solving sessions previously. The group of individuals becomes a team.

The steps in problem solving are similar whether carried out by an individual or a group. An individual always has consensus, but with a group, because everyone is different with different backgrounds and experiences, agreement may not always be possible. However, it is the differences between people that give the group problem-solving process strength—more backgrounds, more experiences, and hence more ideas.

The role of the facilitator is to use the differences in individuals to the gain of the process, while not letting the differences interfere in the process.

9.5.1 Building Consensus

Consensus building starts firstly with an agreement as to the definition and proceeds throughout the process culminating in a solution. Consensus does not mean that senior personnel have already come to a solution and the group is there to rubber-stamp it.

Consensus may involve a compromise, but does not have to. A compromise made in the interests of the process, but not totally wanted by an individual, may struggle later for lack of commitment and ownership.

Time deadlines can be the enemy to establishing consensus. Consensus takes time to build. A time deadline can be used to force an outcome, but the question remains as to the value of such an outcome.

Voting to reach a solution always leaves some people unsatisfied even when aware at the start that a vote is to be taken.

Experts may be used to provide direction on specialist issues and avoid coming up with solutions based on incomplete information. Deferring to an expert is one way that people can save face in an impasse situation.

10

Decision Making with Multiple Objectives

10.1 Introduction

Multiple objectives may come about in synthesis through the incorporation of intangibles (social and environmental* issues), but generally through the desire to have solutions best in many senses, and not just best in cost, for example.

The terms "decision" and "control," when used in the sense of that which is chosen by the engineer, are used interchangeably here.

Outline: The chapter looks at quantitative and qualitative methods for the multiple objective case. Aspects of collective decision making, including the social welfare function (SWF) and public participation in decision making, are treated.

The term "multicriteria" is used by some writers interchangeably with multi-objectives.

10.1.1 Triple Bottom Line

Triple bottom line (TBL) refers to company voluntarily reporting against three matters—financial/economic, social, and environmental. Sustainability and corporate social responsibility (CSR) also look at these three matters. Related to these is ESG (environmental, social, and governance) reporting.

Typically, these three matters are reported separately, though they may be in the one "integrated" document. At the present time, although there is much written on the subject, no one seriously combines all matters in order to be able to compare against some company benchmark or to compare different companies. Ideally, these matters should be interpreted as objectives, and some best company performance determined in a multi-objective way.

---

* The term "environment" in this chapter refers to the natural environment, not the earlier systems definition of "all that is outside the system."

Exercise 10.1

What does TBL mean in environmental impact studies? How does it treat environmental and social intangibles?

TBL reporting recognizes that it is not always appropriate to convert everything into monetary units of measurement and base reporting purely on financial performance. TBL enables a company to report on financial, social, and environmental issues, and utilizes qualitative measures and monetary and nonmonetary quantitative measures to present a picture of a company's performance. Examples of nonmonetary quantitative reporting would be carbon footprint, solid waste volumes, safety statistics, and employee and customer satisfaction survey results. Qualitative reporting examples would be compliance with regulations and guidelines, commitments to sustainable development, and social responsibility.

TBL recognizes the broader stakeholders—employees, shareholders, and the wider community. In this sense, TBL displays some transparency of a company's commitment to the social and environmental impacts of its business, policies, and the initiatives on its stakeholders.

Exercise 10.2

How is TBL reporting dealt with in terms of

1. Accounting standards
2. Reporting for stock exchange purposes (for publicly listed companies)

There is a trend toward greater transparency and accountability in public reporting and communication, reflected in a progression toward more comprehensive disclosure of corporate performance to include the environment, social, and economic dimensions of companies' activities. This trend is being largely driven by stakeholders, who are increasingly demanding information on the approach and performance of companies in managing the environment and social/community impacts of their activities. How this is reported to stakeholders or the stock exchange raises a series of complicated questions.

10.2 Approaches and Examples

Much of synthesis involves dealing with a number of competing objectives. For example, projects often are managed with respect to cost, time, and quality objectives. As another example, designing water resource systems may require four objectives to be considered: national economic development, regional economic development, environmental quality, and other social effects.

10.2.1 Approaches to Dealing with Multiple Objectives

Where there are multiple objectives, they may be *noncommensurate* and *conflicting.* One or more of the objectives may be more dominant than the others.

The treatment of such multi-objective cases involves subjectivity while the treatment of the single objective case does not. Hence, any result in the multi-objective case can always be criticized on this ground.

The three broad ways of dealing with multiple objectives are as follows (Carmichael, 1981, 2004):

Let the objectives by G_1, G_2, ...

1. *Assemble all the objectives into a composite single objective,* for example,

$$G = \alpha_1 G_1 + \alpha_2 G_2 + \cdots$$

   where α_i, i = 1, 2, ... are weighting factors that also take into account the different units of measurement of the objectives. But other means of combining the objectives are possible.

   The subjectivity is introduced in the selection of the weighting factors, in the earlier example.

2. *Isolate one of the objectives as a sole objective and convert the other objectives to constraints* (suitably modified).

   The subjectivity is introduced in the choice of the sole objective, and the constraint levels set on the remainder interpreted as constraints.

3. *Develop an independent best solution for each objective in isolation, and then trade-off the multiple solutions obtained.*

   The subjectivity is introduced in the trading-off process.

People deal with this subjectivity in their own ways. Where more than one person is involved, the situation gets complicated further; how do you deal with multiple different opinions? And where an organization is involved, how is the view of the organization established?

The Analytic Hierarchy Process (AHP), among other methods including "gut feel," can be used to establish the weightings of objectives and subobjectives.

The reinterpretation of some objectives as constraints, which are not to be exceeded or gone below, is attractive to many people's way of thinking.

Evaluation matrices may be used to compare alternatives.

Exercise 10.3

Assume a venture involves community input at the appraisal stage. How do you rank or weight the objectives in such a case?

A developer may be solely interested in profitability, the government may be interested in influencing developments, environmental protection, and the effect on voters, while the public may be interested solely in protecting local concerns. How do you acknowledge all the objectives in such a case?

Can voting overcome such dilemmas? Discuss.

What roles do mediation, negotiation, and bargaining have in resolving such dilemmas?

Exercise 10.4

Consider the three broad approaches for dealing with multiple objectives.

Historically, people have considered money as the important objective in most decision making. Now, with climate change, they are considering a second objective, namely, carbon emissions. So now we have a multiple (two) objective decision scenario. Soon, water and food will be dominant public issues and will need to be included.

Countries are unsure as to whether they should legislate to enable carbon trading or to legislate a carbon tax, or a blend of the two. Both approaches effectively put a price on carbon and convert carbon to a money measure, but in different ways.

The exercise is asking you to compare carbon trading versus a carbon tax in terms of the three broad approaches for dealing with multiple objectives. Do this from a synthesis viewpoint, and not a political, environmental, business, or emotional viewpoint. Do this comparison by looking at the objectives and constraints and how each (that is, carbon trading and carbon tax) deals with the carbon emissions objective, in reducing everything to a single objective related to money.

Example

Assume the synthesis requires both minimum cost and maximum production. Cost has units of money; production has units of output/time. A combined objective may look like

$$(\text{min})\, w_1 \times \text{cost} - w_2 \times \text{production}$$

where w_1 and w_2 are weighting functions subjectively chosen with full knowledge of the different units of measurement involved and reflecting the relative importance of each objective. Alternatively,

$$(\text{min})\, \frac{\text{cost}}{\text{production}}$$

A single objective together with a constraint may look like

$$(\text{min})\, \text{cost}$$

$$(\text{subject to})\, \text{production} \geq \text{specified amount}$$

or

$$(\text{max})\, \text{production}$$

$$(\text{subject to})\, \text{cost} \leq \text{specified amount}$$

10.2.2 Typical Applications

Multi-objective synthesis is dealt with everyday by everyone, possibly without their realizing it. In commercial applications, the most common are

- Equipment and plant selection
- Tender evaluation and assessment
- Project selection/end-product selection
- Employee selection

In personal transactions, there are, for example,

- The selection and purchase of a piece of clothing
- The selection and purchase of a motor vehicle
- The selection of a partner

10.2.3 Selection of Plant and Equipment

The selection of equipment may require consideration of multiple objectives:

- Technical
- Economic/financial
- Emotive

Additionally, equipment selection for public-sector bodies may also be influenced by matters such as politics and government policies.

Economic/financial objectives may be, for example,

- Initial cost
- Life cycle cost
- Cash flow considerations
- Cost/unit of work or time
- Return on investment

Emotive objectives may be, for example,

- Color
- Aesthetics, "looks"
- Patriotism
- Image
- Shape
- Size

Technical objectives may be, for example,

- Procurement time for the equipment (particularly relevant for imported equipment)
- Ability to perform the required task (capability)
- Productivity and utilization
- Efficiency
- Cost per operating time
- Reliability and availability: expenditure necessary to maintain the equipment
- Warranty—terms and conditions, validity period
- Versatility

- Mobility and maneuverability
- Transportation requirements
- Maintenance facilities: available maintenance/repair expertise
- Spare parts availability: procurement time for spare parts (particularly relevant for imported equipment), the cost of spare parts, and the cost of repair
- After sales service
- Compatibility with existing equipment
- Durability (some equipment is viewed as disposable)
- Environmental effects: emissions
- Design life
- Available operator expertise

Utilization is the proportion of time the machine is working (when the machine is available for work), while availability (serviceability) refers to the time when not being maintained or repaired.

Versatility may be an important consideration. A machine that can perform multiple tasks may be preferable to a specialized machine. The versatility improves the chances of utilization of the machine, though its productivity in each task might not necessarily be equal to a machine that only performs a dedicated task.

All selection matters are considered with regard to the work in hand, and in particular with regard to the work's extent, nature, and duration.

For smaller and more diverse work tasks, versatility becomes a more prominent issue. For extensive work of the one type, cost per production becomes the dominant objective.

Equipment working at less than full capacity tends to be not as cost effective as equipment working at full capacity. The temptation to acquire the biggest machine needs to be considered carefully. On the other hand, small machines working at overcapacity may influence the reliability (probability of no failure) of the machines.

Quick access to parts and service is desirable in order that production is little affected by breakdowns. The alternative is to keep a large inventory of parts, but this ties up capital.

Selecting new equipment that is compatible with existing equipment, and allowing interchangeability of parts, can offer some economies in spare parts storage, maintenance, and training personnel (both operators and maintenance staff).

Most equipment, except for that which is highly automated or robots, needs an operator. Some equipment, especially mobile equipment, is not so simple to operate. Accordingly, the operation design features may influence equipment selection. Operation design features include the cabin, seating, location of instruments and controls, work visibility, comfort (including heating and air conditioning), and safety devices. Design influences an operator's effectiveness. Older equipment had limitations with hand controls and lever and pedal mechanisms with low mechanical advantage, leading to fatigue or the need for help. Design also influences the cost and the ease/difficulty with which maintenance and repairs can be made.

Weather and environmental effects—temperature, moisture, wind, air pressure, and a corrosive environment—can influence equipment selection. For example, temperature has a significant effect on the efficiency of an internal combustion engine. Rain or snow creating a wet ground surface will result in poorer traction for equipment, especially equipment with rubber tires.

The current licensing of available operators might be considered. Licensing is dependent on equipment type and capacity, and different licenses may be required for different equipment. If the licenses held by available operators are not appropriate, then training of these operators may be required, and there is the cost of the extra license. A familiar example is that of a car driver's license that only allows the holder of the license to drive cars and small commercial vehicles including small trucks, but not large trucks.

Government bodies, whether through health and safety guidelines or other reason, may require the registration of large equipment (for example, cranes) but not of small equipment.

Exercise 10.5

List any other issues that may affect the technical selection of equipment and rank all issues as they influence the choice of equipment in your workplace.

Exercise 10.6

Given that an operator's cost is (almost) fixed irrespective of the size of a piece of equipment, but that a piece of equipment's production and cost increase with size, does it follow that the optimum piece of equipment is the largest that a single operator can handle? Explain your reasoning.

Cost (monetary constraint) may not be the dominant issue in equipment selection. There may be special project requirements. Time (upper limit) may be a constraint. As well, equipment may be hard to come by in certain geographical areas and at certain times. Consider the following two case examples.

Case Example

In salvaging a ship, power (via a generator) was needed. Some constraints were as follows:

- Local barges for transporting a generator were very small.
- A generator had to be able to run at certain amps and voltage, and on diesel.
- The generator had to be able to be lifted onto the vessel being salvaged.
- The generator had to have self-contained cooling.
- The generator was needed straight away before the vessel sank.

Case Example

Some wharfs have no fixed cranage but have load limits. There may be a need to hire cranage. In some instances, it may be better to hire a barge with cranage than to overload the wharf. It may also be financially attractive to do it this way if an available cranage contractor realizes (or thinks) it has a monopoly and decides to be unreasonable with its pricing.

Typically the multi-objective synthesis is handled through the engineer attaching weights to the various objectives and then assessing each potential piece of equipment against each of the objectives. The weights reflect the relative importance of each objective. Both the weighting and assessment against each objective are subjective processes.

Each potential piece of equipment then receives a total score obtained by multiplying the objective weight by that equipment's assessed mark for the objective and summing over all objectives.

The objectives may be weighted on any scale, for example, from 1 to 10. The assessment against each objective may be done on any scale, for example, from 1 to 10.

Example

Assessment scores (out of 10) for machines A and B against each of five important objectives are given in columns (2) and (4) of Table 10.1. The objectives weightings (out of 10) are given in column (1).

Based on these example numbers, machine B scores higher than machine A and therefore might be preferred.

Other selection methods, or ways of handling the multi-objective synthesis, are available.

TABLE 10.1

Example Equipment Evaluation

Objective (1)	Machine A (2)	(3) = (1) × (2)	Machine B (4)	(5) = (1) × (4)
Reliability (weight 10)	5	50	7	70
Cost (weight 10)	9	90	10	100
Versatility (weight 2)	6	12	4	8
Spare parts (weight 5)	6	30	5	25
Productivity (weight 8)	8	64	8	64
Total		246		267

Exercise 10.7

How else, to this example way, might you deal with multi-objective synthesis?

Exercise 10.8

How does an organization select the weightings for the objectives (as compared with an individual's opinion), and how do you deal with different people in an organization scoring against the objectives differently?

10.2.4 Tender Evaluation and Assessment

Contractors supply requested information that supports their abilities to carry out the tendered work within any time constraints and to any necessary standards. Tender evaluation involves the owner not only looking at the price, but price in conjunction with technical matters and the work and capabilities of the contractor. Attention might be directed at a range of matters including

- Price relative to the owner's budget
- Work method
- The accuracy of calculations
- Unbalanced bid rates

- Alternative offers including special conditions included in the tender and alterations to the contract conditions
- The work program including completion time
- The contractor's past performance with regard to claims, quality, performance, etc.
- Contractor's resources and capacity: equipment
- Proposed site team organization: personnel
- Safety handling
- Specialist abilities
- Quality "control"/assurance practices
- Engineering support by head office
- Other work currently being performed
- Resumes of key staff, training
- Company financial status, share holding, bankers
- Experience, recently completed work
- References and the names of persons or organizations for whom they have worked

The selection of the preferred tenderer is multi-objective synthesis.

Common practice for public sector authorities is that the objectives and weightings are nominated pre-tender, for probity reasons. The transparency associated with prepublishing objectives and weightings removes any ability for discretion or ability to readjust the weightings partway through the tender evaluation process.

Exercise 10.9

An approach, sometimes used on international projects, is to ask tenderers to submit their tenders in two envelopes—the so-called two envelope approach. The first envelope, opened by the organization wanting the work done, is that which contains information about the tenderer's qualifications, experience, personnel, financial backing, quality assurance program, industrial relations program, etc. Tenders are ranked (a subjective activity) based on this information. The second envelope contains the tenderer's price. Tenders are ranked for a second time but now based on their prices.

How do you then select the preferred tenderer? In effect, the two-envelope approach has reduced tender selection to that involving two objectives (nonprice and price) with tenders ranked against each objective.

How do you resolve the potential trouble when the tender ranked "best" against the first envelope information is ranked worst against the second envelope price? Or vice versa?

Exercise 10.10

Suggest a ranking scheme for assessing tenders.

A disciplined approach to tender evaluation is recommended, looking at all the key technical and financial factors in a fair and impartial manner.

Owners may find an apparent conflict between their desire for lowest price and project duration and quality expected.

Different tenders offering different prices and project durations will have to be compared on a proper economics basis. This would take into account, among other things, any benefits to the owner for early project completion (no holding charges, rental, taxation implications, …) Other situations where tenderers alter conditions of contract differently or where, say, the rise and fall conditions are different, make tenders not so readily comparable; this also applies generally when comparing conforming (complying) and nonconforming (alternative) tenders.

Exercise 10.11

Part of the conditions of tendering for one job stated: "The objectives… may not be accorded equal weight." The objectives listed were as follows:

- Tender price.
- Compliance with the technical requirements.
- Previous experience and performance.
- Compliance with conditions of contract.
- Compliance with quality assurance provisions.
- Proposed program of work and work sequence.
- Should the tenderers be informed of the weighting? If yes, should the units of the weightings be given to the tenderers? How is the noncommensurability (different units) to be treated?

Should the tenderers be informed of how each of these objectives is to be measured? Are scales of measurement being used?

In the evaluation of tenders, if more than one individual is involved, how are individual evaluations weighted?

How subjective is the process of tender evaluation, given weightings for the different objectives and given that more than one individual may be involved in the evaluation?

Example

In one organization, the tender evaluation committee is never less than three persons. The committee meets (prior to the opening of tenders) to establish the evaluation methodology. Here the mandatory, technical, and pricing weightings are determined. The committee notes that it will need to do a sensitivity study to determine the winning tender if two or more bids are within 5% of each other in the final assessment. The procedure is set up in a manner that each committee member (once the tenders are open) assesses the tenders individually and has his/her own spreadsheet to insert the results. Once all members have assessed the tenders, their individual results are then inserted into a common spreadsheet, and the combined totals give the final results. The committee then agrees, recommends, and signs off on this result. This procedure allows an unbiased and fair approach and does not allow the project owner to manipulate the final result. However, this procedure does have its downside, in that it is not unusual for the tender, considered least likely, to win at times.

Exercise 10.12

One way to obtain scores based on peoples' different ideologies and mind-sets is with a Delphi-style technique, where all people involved give their opinion, and once all opinions are collated, team members are given the opportunity to change their scores. There have been many books written on group decision making. Group decision making can be flawed in some cases and will ultimately depend on, for example,

- Composition of the group (heterogeneity of experience and personalities within the group)
- Interrelationships between group members
- Internal politics and agendas
- Degree of power held by each team member (the lowest of the hierarchy, although maybe having the best idea, may be ignored)
- Behavior of the group leader

A lot of the issues in the list relate to power and internal politics and agendas. Unfortunately, this can translate to the "whims of senior staff", which could be to "get the lowest price, no matter what." This may be an extreme or cynical view, but this is commonly the case, even though it may be disheartening and, in some cases, disillusioning.

How do you remove the people influence from tender evaluation?

Exercise 10.13

There are three basic ways of dealing with the multi-objective case, as outlined earlier:

1. Combine all objectives into a single objective.
2. Select one objective as the main objective and convert the remaining objectives into constraints.
3. Evaluate according to each objective separately and trade off the resulting choices.

The first method may be used by owners through using weightings for the objectives. The second method is used in principle when the owner specifies some objectives as being mandatory. Why is not the third method used by owners?

Exercise 10.14

Commonly, conforming (complying) tenders are requested by owners, and provided this is done, "... alternative tenders for any section of the work will be considered but may not be accepted."

Given that you are now comparing apples with oranges (that is, conforming and nonconforming tenders), how does the evaluation according to weighted objectives now work? Does this not introduce more subjectivity into an already subjective approach? How can you fairly evaluate tenders under such circumstances?

One way, one owner uses to compare conforming and nonconforming (alternative) tenders is to use a table such as the following:

	Tenderer A ($)	Tenderer B ($)
1. Brief details of alternative tenders		
2. Tender prices as submitted		
3. Errors in tender prices/rates authorized for correction; mathematical corrections		
4. Commercial adjustments to comply with specification		
5. Technical adjustments to comply with specification		
6. Tender prices adjusted to comply with specification		

	Tenderer A ($)	Tenderer B ($)
7. Adjustments for evaluation purposes only		
—Preference		
—Other		
8. Comparison of adjusted prices		
9. Order of tenders after adjustment		

How successful might it be in converting differences due to alternative tenders (compared with conforming tenders) to dollars?

Exercise 10.15: Example Tender Evaluation

Outline of the constraint, objectives, and their weighting:

Mandatory Constraint

0. Site inspection

General (Nonprice) Objectives

Objective	Max Score
1. Quality assurance	10
2. Time frame and program	30
3. Experience	20
4. Resources	30
5. Compliance with conditions of contract	10
Total general score	100
Normalized general score	100

Commercial (Price) Objective

Objective	Max Score
Total evaluated price ($)	
Normalized price score	100

Total Weighting Summary

Objective	Max Score
Normalized general score	60
Normalized price score	40
Combined score	100

Example entries in evaluation tables:

Evaluation Matrix

Objective	Evaluation Method	Weighting Factor n	Weighting Factor n.n	Tenderer A Score/10 for All Items	Tenderer B Score/10 for All Items
1. Quality assurance		1			
1.1. Does the tenderer have a Quality System in place?	Score 10 points for evidence of accredited QA system, 4 for evidence of a system without accreditation, and 0 for no system		1	10	10
Subtotal points		10		10.0 (=10/10 × 10)	10.0 (=10/10 × 10)
2. Time frame and program		3			
2.1. Does tenderer's program meet required time frame?	Score 10 points for tender with shortest program. More than 1 week after required deadline scores 0 points		1	6	7
2.2. Does the program consider all aspects of the proposed work?	Evaluate extent of details as defined in the general evaluation table		1	6	7

Objective	Evaluation Method	Weighting Factor n	Weighting Factor n.n	Tenderer A Score/10 for All Items	Tenderer B Score/10 for All Items
2.3. What is the tenderer's performance in meeting programmed deadlines?	Score 10 points for evidence of previous satisfactory delivery performance, 5 points for average or no experience, and 0 points for poor performance		2	8	2
2.4. Does the tenderer have the capacity to complete the work?	Score 10 points if requirements can be met, 0 points if not		1	8	8
Subtotal points		30		21.6 (=30/50 × 36)	15.6 (=30/50 × 26)
3. Experience		2			
3.1. As required by specification, has the tenderer supplied a list of references supporting their performance?	Score 2 points for each supportive reference (to be checked), maximum 10 points		3	10	7
3.2. Does the listed experience address similar projects?	Evaluate as defined in the general evaluation table		1	5	8
3.3. Have any proposed subcontractors demonstrated suitable experience on this type of work?	Evaluate as defined in the general evaluation table		1	5	5
Subtotal points		20		16.0 (=20/50 × 40)	13.6 (=20/50 × 34)

Objective	Evaluation Method	Weighting Factor n	Weighting Factor n.n	Tenderer A Score/10 for All Items	Tenderer B Score/10 for All Items
4. Resources	3				
4.1. Are the key personnel suitably qualified for the proposed project?			1	6	6
4.2. Has the tenderer demonstrated a commitment to resourcing the project sufficiently to achieve program?			1	6	3
4.3. Is construction staff suitably qualified or equipped to participate in the proposed project?			1	6	6
4.4. What proportion of the work is to be completed by subcontractors?			3	6	2
4.5. Demonstrated a capacity to resource and achieve their program?			1	6	4
Subtotal points		30		18.0 (=30/70 × 42)	10.7 (=30/70 × 25)
5. Compliance with conditions of contract	1				

Objective	Evaluation Method	Weighting Factor n	Weighting Factor n.n	Tenderer A Score/10 for All Items	Tenderer B Score/10 for All Items
5.1. Does the tenderer comply with the conditions of contract			1	10	10
Subtotal points		10		10.0 (=10/10 × 10)	10.0 (=10/10 × 10)
Total point score—general		100		75.6	59.9

Weighting Factors

0. Requirement not applicable
1. Desirable requirement
2–5. Important (select value for relative importance)

Compliance Evaluation Table

0. Requirement not met
2–4. Partial compliance, but meets service requirements
6. Requirements fully met
7–10. Tangible value to owner in excess of requirements

Mandatory Constraint

	Tenderer A	Tenderer B
0. Site inspection	Yes	Yes

General (Nonprice) Objectives

Objective	Max Score	Tenderer A	Tenderer B
1. Quality assurance	10	10.0	10.0
2. Time frame and program	30	21.6	15.6
3. Experience	20	16.0	13.6
4. Resources	30	18.0	10.7
5. Compliance with conditions of contract	10	10.0	10.0
Total general score	100	76	60
Normalized general score	100	100.0	79.3

Commercial (Price) Objective

Objective	Max Score	Tenderer A	Tenderer B
Total evaluated price ($)		560 k	720 k
Normalized price score	100	100.0	77.9

Total Weighting Summary

Objective	Max Score	Tenderer A	Tenderer B
Normalized general score	60	60.0	47.6
Normalized price score	40	40.0	31.2
Combined score	100	100	79

What would be the basis for choosing the weightings of 10, 30, 20, 30, and 10 for the general (nonprice) objectives? How sensitive is the result to changes in these weightings? How much do you have to change these weightings or scores before Tenderer B becomes as attractive as Tenderer A?

What would be the basis for choosing a 60:40 split between the general (nonprice) objectives and commercial (price) objective? Different weightings can produce different outcomes. Is the technical:commercial split chosen to reflect the nature of the work? Would the split be decided before opening tenders or after opening tenders? How might probity issues be involved here in the choice of the split, given that different evaluation outcomes are possible with different splits?

Why normalize separately with respect to general (nonprice) objectives and commercial (price) objective? Is this for internal consistency within the general (nonprice) objectives and the commercial (price) objective, acknowledging that the two are like an apple and an orange?

Where do the n.n weights come from? How sensitive could the result be to changes in the n.n weights?

10.2.5 Project Selection/End-Product Selection

In project work, there is always a choice between different ways of doing things and choice between different end-products. The selection of a preferred end-product and selection of a preferred project are undertaken during project pre-feasibility and feasibility studies (Carmichael, 2004). Both financial and nonfinancial matters are involved.

Ventures (including project and end-product) generally require justification on economic grounds. Accordingly, an economic feasibility study identifies the benefits and costs. Future benefits and future costs are commonly discounted to present-day values.

Economic justification not only requires that the benefits exceed the costs but that the venture compares favorably with the status quo, along with alternative proposals.

Some of the general information needed for economic feasibility studies is

- Total investment costs
- On-going end-product costs, income
- Cash flow
- Financing
- Financial appraisal using discounted sums: cost–benefit analysis
- National economic issues including exchange rates

The economic feasibility study recognizes that many ventures require the expenditure of large amounts of money to design, put in place and to operate.

Considerations other than financial may play a role in feasibility studies. These include

- Technical, regarding constructability and function
- Legal, including approvals
- Environmental, both statutory and community viewpoints
- Social impacts on the community
- Sustainability and its effect on future generations
- Political at all levels of government

Example

Consider the issues associated with the selection of the route that a road might take between two locations (origin/destination), A and B (Figure 10.1).

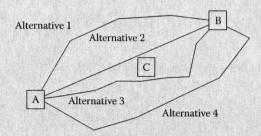

FIGURE 10.1
Example, possible route alternatives between A and B.

The different route alternatives would be assessed with respect to, for example,

- Costs (resumption of properties, construction, maintenance, …)
- Benefits (reduced travel time, reduced accidents, …)
- Environmental issues (noise, pollution, impact on flora, fauna, heritage and indigenous sites, …)

This is made more complicated because of the different groups involved:

- The road authority
- The community at large, the majority of whom may never use the road, or even be interested in the road
- The communities at the route ends, A and B, which benefit from decreased travel times
- The community along the route, C, which is subjected to increased noise and pollution
- Special interest groups concerned with, for example, flora, fauna, heritage, and indigenous sites

Each group will come up with a different solution based on the multiple objectives.

10.2.6 Personal Applications

10.2.6.1 Vehicle Selection

For personal vehicle selection, in addition to the earlier listed objectives for equipment and plant selection, there are also the image objectives, which are targeted by marketing (promotional and sales) personnel:

- Color
- Shape and style
- Brand
- Age

For some people, these are the dominant objectives, and functional objectives, such as reliability and spare parts availability, do not even get considered.

10.2.6.2 Clothing Selection

Fashion may dictate that image objectives (color, pattern, style, and brand) are the only objectives considered. For clothing purchased for utilitarian reasons, the objectives are quite different and relate to initial cost, lifetime usage, and maintenance considerations.

10.2.6.3 Partner Selection

When a person "chooses" a partner, there are a number of objectives (traits), which are being balanced in that person's head at any one time. These include

- Physical appearance—height, weight, shape, looks
- Age
- Value system—beliefs, likes/dislikes
- Compatibility of interests—social, work, sport, humor
- Background—education, upbringing, experiences
- Personality

Women may choose differently to men. For example, there is a view that the most important traits to a woman are economic resources; good financial prospects; high social status; older age; ambition and industriousness; dependability and stability; athletic prowess; good health; love; humor; and willingness to invest in children. For men, there is a view that the most important traits are youth; physical attractiveness; and particular body shape/proportions (waist, hip, leg, height, etc., ratios). Intelligence and kindness might also be considered desirable traits.

Objectives weightings may get adjusted during the search process when Mr. Right or Ms. Right cannot be found, or the search process seems to be never coming to a conclusion.

Women may become less selective in their favored traits as the operational sex ratios decrease (more women relative to men) and become more selective as they increase (more men relative to women), and vice versa for men. This is a supply-and-demand scenario and hence possibly the reason for the term "marriage market." If there are more men than women, then the "supply" of men is high, their "value" decreases, the "price" or purchasing power men acquire plummets, and as a result men have to lower their standards and become less selective in order to find a partner. Something similar happens to women.

As couples progress in time together and mature, weightings change and scores against the objectives change.

10.2.7 Employee Recruitment

The employee recruitment process carried out in organizations follows a similar multi-objective format. Example objectives include appearance, dress, personal and conversational manner, work experience, education, and attitude. Weightings might change depending on the urgency of an organization's needs and the supply in the labor market; recruiters may accept previously considered unacceptable people in a tight labor market. An example is where a company now employs a person, who has had his/her vehicle driving license revoked by the state, to drive an off-road truck. This is legal because the truck is operating on private land. Previously, losing a license because of driving offences may not have been tolerated because it would represent a lack of discipline and ability or desire to obey instructions. Another example is where a person with a criminal record is now employed where previously they may not have been.

The saying, "hire attitude over aptitude," reflects a higher weighting being given to the former.

The competencies that are associated with a job's description are used as constraints—the candidate either has the competencies or not.

10.2.7.1 Needs and Wants

In human resource selection, competencies might be labeled "needs" or "required," or "essential," while the objectives might be labeled "wants" or "desired" or "desirable." The "wants" assume different weightings in different situations. An example is that of a truck driver. "Needs" are the ability to drive a truck. "Wants" are experience, license, training, good personality, and so on.

But in any selection, the "needs" can be separated from the "wants." Needs are constraints. Wants get incorporated into the objective function.

Sometimes people think that things with high priority (and thus high weightings) are essential, but this is incorrect. A better way to describe them is as "having a large effect on the outcome" or being "very important in order to achieve success" or similar. If the large weighting is stripped away (because of changed circumstances), then they become not as important. However, a true constraint (a need) is always required. Desperate circumstances in human resource recruitment often expose the true "needs" as the bare minimum.

Exercise 10.16

How realistic are we when we choose weightings? Can these weightings be substantiated?

Exercise 10.17

Frequently, you see published rankings of universities around the world. These are interesting examples involving multiple objectives. People are asked to score universities against multiple objectives. The promoter of the ranking exercise then combines the scores using unpublished weightings for the objectives.

Critically appraise such undertakings by considering the indefiniteness and subjectivity involved in

- The possible inappropriateness of the questions asked and their connotations
- The sample used to provide replies
- The scoring scale used for each objective
- The weighting chosen for each objective

Generally, you will find that by appropriate choice of questions, sample selection, scoring scales, and weightings, you can arrange for any university of your choosing to be crowned number one in the world. So why do supposedly intelligent university people (other than those at weaker universities) give such rankings any credence?

Exercise 10.18

CSR, sustainability, and TBL reporting, in common, pay attention to not only economics/finances but also the environment and social issues. If you think of these as objectives, then you have multi-objective synthesis whenever a decision is being made (here "multi" = 3, but each objective has many subobjectives).

Now all objectives and subobjectives have different units of measurement: economics/finances in terms of dollars; the environment in terms of pollution, noise, flora, fauna, ...; and social in terms of well-being, safety, employment, ...

An economist will tell you that all environmental and social issues have a dollar value, but this view has many opponents. Hence, you are stuck with multi-objective synthesis where the objectives have difficulty being combined under a single unit of measurement.

So the exercise is asking you to articulate how organizations view synthesis while addressing economic/finance, environment, and social issues simultaneously.

10.2.8 Noninferior Solutions

It can be very difficult to balance the relative worth of different objectives in practice and determine a single optimum that will be acceptable to all concerned. It may be more preferable to think in terms of inferior and non-inferior solutions.

Inferior solutions are dominated by other solutions that are preferable on all counts. For noninferior solutions, there is no solution that is clearly better for all objectives. These are a subset of all admissible solutions, perhaps worthy of further scrutiny, and from which a preferred solution might be obtained.

> **Exercise 10.19**
>
> On what basis might you say that a solution is better than, or as good as, another solution in terms of all the objectives?

10.3 Collective Decision Making

In much synthesis, a number of groups with different interests are likely to be affected. A best overall decision for all groups may not be able to be found. This does not mean that mathematical techniques cannot be used, but that resulting solutions are unlikely to be acceptable to all interested groups.

Some of the techniques used in the area of collective decision making are listed in the following text.

10.3.1 Social Welfare Function

The social welfare function (SWF) can be thought of as a kind of expression of the utilities of different individuals or groups U_i:

$$SWF = f(U_1, ..., U_n)$$

There are concerns in determining the form of the function f. If society valued every individual equally, a summation or a weighted SWF may be satisfactory. Others argue that the SWF should give greater emphasis to ensuring that everyone achieves some minimal level of satisfaction and give less emphasis to individual levels of utility.

> **Exercise 10.20**
>
> How do you establish the utility of a group of people, as opposed to an individual?

10.3.2 Public Involvement

Public involvement in decision making has increased markedly in recent years. It is now extremely difficult to undertake any venture without some part of the community becoming involved. Groups with diverse interests are now organized within the community and seek to influence decision making (especially with respect to public infrastructure) to ensure their requirements are met.

Venture proponents are now forced to deal with the public as part of synthesis not only because of legislative requirements but also because of the power of particular lobby groups. This means that decision makers need to be adept at dealing with community groups and listening and acting on public concerns.

The aim from the decision maker's point of view should be to obtain a consensus with the public as to which venture should proceed. The participants from the general public are likely to be aiming for protection or furtherance of particular interests. There is a range of methods of incorporating public participation.

The major costs of public involvement in the decision making are in setting up and attending meetings and providing information, as well as the extra time required in reaching a decision.

The conflicts that arise out of public involvement need to be handled well for an effective decision to result from the process. For this reason, the negotiation and bargaining are important to the outcome.

> **Exercise 10.21**
>
> Should opinions of different community groups be weighted or treated equally?

11

Optimization

11.1 Introduction

11.1.1 Design

Established design procedures are predominantly iterative in nature. The iterations arise from the analysis-based mode of attack on design and are not inherent in design. Optimization and related techniques eliminate much of the iterative process.

The iterative design process is typically structured through the steps of

- Definition
- Objectives and constraints statement
- Alternatives generation
- Analysis and evaluation
- Selection

Feedback occurs in an effort to refine the process at any of the steps.

Using optimization approaches, the iterative design process is short-circuited. The approaches enable a direct path to the "selection" step, effectively jumping over the intermediate steps. The "alternatives generation," "analysis and evaluation," and "selection" steps merge as one. Other than that, optimization approaches hold no mysteries.

Optimization is a technique from the field of operations research and systems engineering. It is an aid in synthesis and problem-solving processes. The range of synthesis cases capable of being handled by optimization is very diverse, but they tend to be smaller in scale than those attainable using synthesis via iterative analysis. The level of mathematical sophistication in some optimization theory can be high, though prospective users need only master a few essential concepts. The chapter demonstrates how to recognize the components of an optimization and set up optimization.

11.1.2 Outline

Deterministic cases only are considered, because it is believed that the returns from considering probabilistic formulations do not match the extra work involved. Also, continuous variables only are considered, because again, it is believed that the returns from considering integer formulations do not match the extra work involved. Where probabilities or integers are involved, some conversion will be required back to the deterministic continuous variable case.

Numerous examples are given to help understand optimization and the formulation of its components.

11.1.3 Optimization Techniques

To fully understand the complete picture of optimization requires a solid foundation in linear algebra and calculus. Numerous optimization techniques have been proposed. There is no one technique that is suitable for all possible situations.

Some of the more common techniques are presented later, but it is by no means a complete presentation. The presentation is also given without complete mathematical rigor. The development proceeds from the elementary to the advanced.

In summary form, the most popular optimization techniques and their range of usage is as follows:

Optimization Technique	Usage
Calculus of extrema (minima, maxima)	Elementary optimization involving algebraic expressions for the synthesis components. Preference is for unconstrained versions
Mathematical programming; this includes linear programming and the many forms of nonlinear programming	The same as for calculus, but where a closed-form result is not possible or where constraints complicate the calculus, a numerical result is obtained
Calculus of variations, dynamic programming, Pontryagin's principle	Dynamic systems, continuous or staged
Numerical evaluation, numerical optimization	A number of possible solutions are guessed, the objective evaluated numerically, and the best solution is chosen. Commonly used by practitioners

Note that not all that can be formulated as optimization can be dealt with by available optimization techniques. Available optimization techniques have a restricted range of applicability. Analysis, on the other hand, usually leads to a unique result. For this reason, synthesis is predominantly looked at in terms of iterative analysis rather than by, say, formal direct optimization techniques. As well, even if an optimization technique is available for a particular formulation, many people may well choose to deal with it in an iterative-analysis fashion anyway or by numerical evaluation.

11.1.4 Packages

There are a number of commercially available computer packages applicable to standard optimization formulations.

Outline: The chapter firstly shows how conventional design is carried out, by means of an example, which is then carried through to later sections to show how optimization works by comparison. Section 11.3 defines what constitutes a general optimization form, from which simplifications and specializations can be made. Included in this section is a discussion on the fundamental types of variables found in the general optimization form. Standard optimization forms are then defined in the next section in terms of the identified components and variables. The next three sections outline more mathematically oriented approaches to optimization. Section 11.8 outlines some of the full scope of optimization forms and theory.

11.2 Conventional Design

11.2.1 Example

An example is used to demonstrate conventional design practices. Consider a shovel-truck operation involving a single shovel and known haul and return routes (Figure 11.1) where the optimal number of trucks is required. The operation is the system from the viewpoint of the optimization.

The behavior of the operation (control–state relation) may be described using queuing equations, numerical simulation, or conventional earthmoving methods; all of these are valid models or representations of shovel-truck operations. All should predict behavior similar to that given in Figure 11.2.

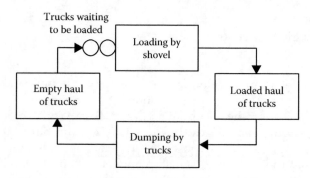

FIGURE 11.1
Earthmoving, quarrying, surface mining operation.

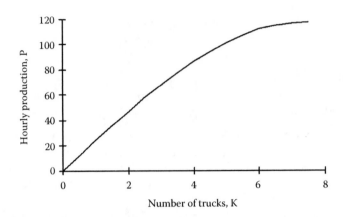

FIGURE 11.2
Approximate model of earthmoving behavior.

There may be constraints on the operation in the form of time limitations: a certain amount of earth has to be moved in a given time period.

The design must be carried out with respect to some objective. A reasonable objective for earthmoving operations is one of (minimizing) the cost per production.

11.2.2 Quantitative Formulation of Design

Consider how this design may be interpreted mathematically. Using the results of queuing theory, simulation, or conventional earthmoving calculations, let the relationship between hourly production and the number of trucks be given by

$$P = 24\left[K - \frac{K^3}{160} \right]$$

where
  P is the hourly production (m³/h)
  K is the number of trucks

This is a crude approximation for the hourly production versus number of trucks behavior (for a given earthmoving operation for K up to about 7 involving loading times of about 5 min, truck capacities of about 10 m³, and a haul, dump, and return time of about 20 min) but will suffice for the present explanation purposes. More exact models could be used. Note also that the earlier relationship between P and K is not restricted to integer values for K; however, in practice, only integer K values are possible.

Let there be a time limitation constraint that the operation has to move, say, 155,000 m³ in a period of 1,500 h. That is,

$$1,500\,P \geq 155,000$$

or

$$P \geq 103.3$$

For an objective function of cost per production, introduce the following notation:

C_1 hourly owning and operating cost of the shovel
C_2 hourly owning and operating cost of a truck

Then the total hourly cost of the operation is

$$C_1 + KC_2$$

The objective function, denoted J, of cost/production becomes,

$$J = \frac{C_1 + KC_2}{P}$$

Formally, the mathematics of the optimization is written as,

$$\min_{K} J = \frac{C_1 + KC_2}{P}$$

$$\text{subject to} \quad P \geq 103.3$$

$$P = 24\left[K - \frac{K^3}{160}\right]$$

$$P \geq 0,\ K \geq 0$$

The last constraints are termed *nonnegativity constraints* and ensure that any result obtained will have nonnegative variables. They are not needed if the calculations are being done by hand, but may be needed if the calculations are being done on a computer.

11.2.3 Conventional Design Approach

Consider how the optimum may be obtained by conventional design. Let $C_1 = C_2 = 100$.

The formulation may be written mathematically as

$$\min_{K} J = \frac{100 + 100\,K}{P}$$

subject to $P \geq 103.3$

$$P = 24\left[K - \frac{K^3}{160}\right]$$

$$P \geq 0,\ K \geq 0$$

Conventional design would proceed as follows:

Step 1: Guess a value for K (the control variable).
Step 2: Calculate the production P (the state variable), objective function J, and the time-production.
Step 3: Check to see that the time-production constraint is satisfied; if the constraint is not satisfied, reject the K value guessed.
Step 4: Continue to return to Step 1 (that is, guessing new values for the control variable K) and continue to work through the steps until a smaller value of J cannot be found.
Step 5: The K value corresponding to the lowest value of J is regarded as the "best" value.

That is, conventional design is essentially an *iterative-analysis* approach.

11.2.4 Alternative Computations

The alternative (via numerical evaluation) is to set up a table or spreadsheet that does the same thing. Table 11.1 has columns of K (the control variable), P (the state variable), and J (the objective function). The information in the

TABLE 11.1

Cost/Production versus Truck Number

K (Trucks)	P (m³/h)	J ($/m³)	T (h)
1	23.9	8.37	6485
2	46.8	6.41	3312
3	67.9	5.89	2283
4	86.4	5.79	1794
5	101.3	5.92	1530
6	111.6	6.27	1389
7	116.5	6.87	1330

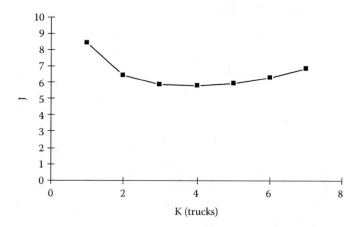

FIGURE 11.3
Cost/production versus truck number.

first three columns of Table 11.1 is plotted in Figure 11.3. The optimum for the case without the time-production constraint (the *unconstrained case*) clearly lies at K = 4. In practice, a fleet of five trucks may be used to allow for maintenance, multitasking, etc., of the vehicles. Only integer K values are used in the computations.

The result with the time-production constraint (the *constrained case*) can be seen from the second column of Table 11.1. The value P = 103.3 lies between K = 5 and K = 6, and so the least cost *admissible* value (that is, a value that satisfies any constraints) is K = 6.

11.2.5 Formulation in Terms of T

Consider a different (and longer) approach to the constrained case. The time, T, to move 1,55,000 m³ is given by

$$T = \frac{1,55,000}{P}$$

where
   P is the hourly production (m³/h)
   T is in hours

T is calculated in the fourth column of Table 11.1. Figure 11.4 plots J versus T. Superimposed on Figure 11.4 is the line T = 1500. Values to the right of this line are *inadmissible* (that is, they do not satisfy any constraints present). The optimum value is K = 6, as before.

Conventional design when carried out systematically in the fashion of Table 11.1 might be referred to as *numerical optimization* or evaluation.

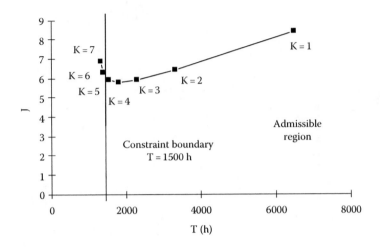

FIGURE 11.4
Cost/production versus time taken.

True optimization conceptually replaces the iterative-analysis search for the best values of the variables by going directly to the best values. (However, be aware that some optimization algorithms may still do iterative calculations, which in many cases give an equivalence with synthesis via iterative analysis.)

Exercise 11.1

Project costs may be divided into direct costs (costs that vary in direct proportion to the project activity and include labor and materials) and indirect costs (costs that generally vary with time or are one-off such as overheads, profits, and running expenses). Indirect costs are estimated to be $3000 per day. Direct costs are estimated to be the following:

Project Duration	Direct Cost
75	75,000
70	80,000
65	90,000
60	100,000
55	120,000
50	150,000

Calculate the project duration of least total (direct plus indirect) cost.

11.3 Components of Optimization

11.3.1 Relationship to the Design Process

Optimization works with (possibly removes the iterations from or assumes that they remain unchanged) the later steps of synthesis via iterative analysis, namely, the steps of

- Alternatives generation
- Analysis and evaluation
- Selection

Optimization approaches are quantitative in nature and so the information considered in these steps (model, constraints, and objective) needs to be put into a mathematical form. Any alternatives generated need to obey a relevant model of behavior and fit within any constraints placed upon the designer.

Optimization, in general, is written mathematically in terms of three components:

- A *model* (representation) of the system or operation expressing the behavior or functioning of the system or operation: a control–state relation
- *Constraints* restricting the number of possible values of the variables
- An *objective function* by which alternative values may be compared

Example

Consider the shovel-truck operation of Figure 11.1. The control variable is K; the state variable is P. The optimization formulation was established earlier as

$$\min_{K} J = \frac{C_1 + K C_2}{P}$$

$$\text{subject to} \quad 1,500P \geq 155,000$$

$$P = 24 \left[K - \frac{K^3}{160} \right]$$

$$P \geq 0, K \geq 0$$

Here, the *system model* is

$$P = 24\left[K - \frac{K^3}{160}\right]$$

The *constraints* are

1,500 P ≥ 155,000
P ≥ 0, K ≥ 0

The *objective function* is

$$\min_{K} \ J = \frac{C_1 + KC_2}{P}$$

11.3.2 Alternative Terminology

Some writers combine the model and constraints, and collectively refer to both as constraints.

Some writers refer to the totality of these three components as a "model of optimization," a "model of the optimization problem," or a "model of the decision-making process"; this use of the term "model" is not followed here because, although not incorrect, it tends to confuse with the earlier defined use of the term "model." These writers get away with such terminology because the term "system" is sufficiently broad to allow it; by calling a system anything they like, then so they have a model of anything they like.

The term *restraint* is sometimes used by writers instead of constraint.

The term "objective function" is perhaps the most popular to denote that entity by which the best value is chosen. However, different disciplines may use different terms such as *optimality criterion, performance index, performance measure, payoff function, figure of merit, merit function, goal, cost function, design index, target function, criterion, mission,* and *aim* (Carmichael, 1981).

11.3.3 Possible Approaches

There are several ways of approaching the example shovel-truck optimization. For now, a graphical approach is given; closed-form and numerical approaches are discussed later.

Consider firstly the case without the inequality time-production constraint. Figure 11.5 shows the system model/equation plotted on a graph with axes comprising the variables P and K, optimum values of which are

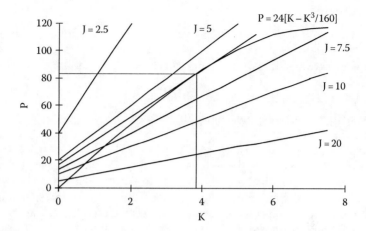

FIGURE 11.5
System model and objective function contours.

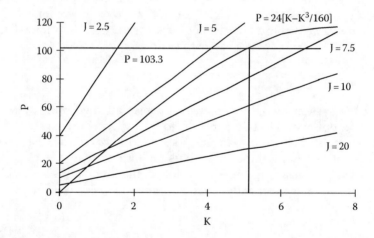

FIGURE 11.6
System model, objective function contours, and time-production constraint.

to be determined. Superimposed on Figure 11.5 are contours of J (for $C_1 = C_2 = 100$). The optimum value is that which gives the lowest J while satisfying the system equation. That is, the tangent (being a contour of J) to the system equation is sought. This occurs at a K value of about 3.9.

Consider now including the inequality time-production constraint (Figure 11.6). The value of K that minimizes J while satisfying all constraints and the system equation is that which lies at the intersection of the system equation and the equality part of the time-production constraint. The approach in this case is made easier by the inclusion of the time-production constraint. This occurs at a K value of about 5.2.

11.3.4 Variables

The earlier example illustrates most of the salient points involved in optimization. Not only does it demonstrate the three fundamental components but it also demonstrates the two fundamental types of variables:

- *Control* (design, decision, input) *variables*
- *State* (behavior, response, output) *variables*

Control variables are those that the designer/planner/manager has direct influence over. By varying the values taken by control variables, the system behavior is changed. In the earlier example, K (the truck fleet size) is the control variable. Behavior or state variables take their values through the system model once the control variable values are established. In the earlier example, P (the hourly production) is the state variable.

Some writers put all variables into one basket and call them "decision variables"; such terminology is not followed here, although it is found convenient at times to not differentiate between the different types of variables. In this book, the terms "decision" and "control" are used interchangeably, and mean that which is chosen by the engineer in synthesis. Optimization and decision making done optimally are the same thing.

11.3.5 Admissible Regions

Calling the system model a constraint enables all the optimization components except the objective function to be called constraints. Values of the variables satisfying these constraints are termed *admissible*. Values of the variables not satisfying these constraints are termed inadmissible. The collection of values satisfying the constraints defines an *admissible region*. The admissible region is independent of the form of the objective function.

Exercise 11.2

Use a diagram for the following. A manufacturer can produce two types of components, which for simplicity may be denoted component 1 and component 2. For component 1, there is the potential for producing 150 per day and the work requirement is 10 people days per 100 components produced. For component 2, there is the potential for producing 250 per day and the work requirement is 5 people days per 100 components.

Profit on component 1 is $800 per 100 and on component 2 is $500 per 100. The manufacturer has a workforce of 20.

Calculate the daily production of each of the components in order to maximize the profit of the manufacturer.

11.4 Standard Forms

A number of standard optimization forms can be identified. The earlier example can be shown to fit within a standard optimization form (SF1 given in the following text). Optimization for the so-called static case is dealt with here; later mention is made of the dynamic case.

11.4.1 Standard Form SF1

Let $x = (x_1, ..., x_n)^T$ be a vector of state variables and let $u = (u_1, ..., u_r)^T$ be a vector of control variables. A standard optimization form (denoted Standard Form SF1, here) becomes

$$\min J = G(x, u)$$
$$\text{subject to } h_j(x, u) \leq 0 \quad j = 1, ..., m \tag{SF1}$$

J is the objective function. G is a nonlinear scalar function of the arguments shown. h_j, $j = 1, ..., m$ are nonlinear scalar functions and include both the system model (equalities) and the constraints (equalities and inequalities). The system model may be thought of as an equality constraint. h_j includes any *nonnegativity conditions* on the variables; these are constraints ensuring that the variables do not take negative values. The inequalities are written as $\leq$ without loss of generality, because $\geq$ type inequalities and equalities may be converted to $\leq$ inequalities, as is demonstrated later. Minimization is also assumed without loss of generality, because maximizations may be converted to minimizations, as is also demonstrated later.

Standard Form SF1 is referred to as a *mathematical programming* form or, if the components are linear, as a *linear programming* (LP) form.

11.4.2 Standard Form SF2

In many situations, it is mathematically convenient to make no distinction between control and state variables. Let $x = (x_1, ..., x_n)^T$ be both the control and state variables. (An enlarged value for n is implied here.) Then Standard Form SF2 becomes

$$\min J = G(x)$$
$$\text{subject to } h_j(x) \leq 0 \quad j = 1, ..., m \tag{SF2}$$

In the n = 2 case, these expressions may look like the example in Figure 11.7.

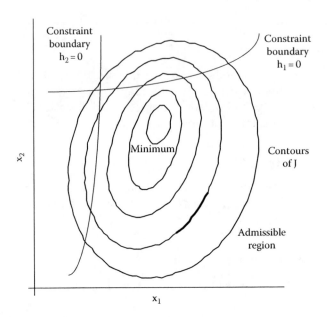

FIGURE 11.7
Example graphical representation of Standard Form SF2.

11.4.3 Standard Form SF3

Where both G and h_j are linear functions, the formulation is referred to as a linear programming form

$$\min J = c^T x$$

(SF3)

$$\text{subject to } Ax - b \leq 0$$

where
   A is an m × n matrix of constants
   b is an m-dimensional vector of constants
   c is an n-dimensional vector of constants

In the n = 2 case, these expressions may look like the example in Figure 11.8.

11.4.4 Conversions between Standard Forms

The standard forms of optimization are given in (SF1), (SF2), and (SF3). Formulations not belonging to these standard forms at the outset may generally be converted to them through various devices.

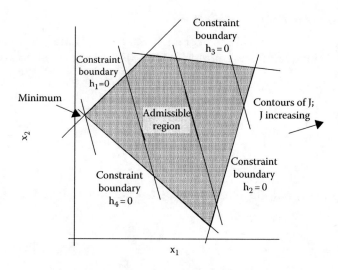

FIGURE 11.8
Example graphical representation of Standard Form SF3.

11.4.4.1 Negative Variables

Where nonnegativity of variables is required such as in the linear programming case, any negative variable can be expressed as the difference between two nonnegative variables.

11.4.4.2 Maximization

The maximization of a function $G'(x)$ is mathematically equivalent to the minimization of a function $-G'(x)$. See Figure 11.9.

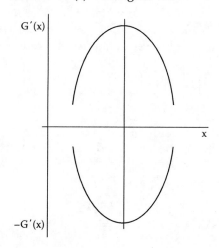

FIGURE 11.9
Equivalence of maximization and minimization.

11.4.4.3 Greater Than Inequalities

Constraints expressed as greater than (>) inequalities may be converted to less than (<) inequalities by multiplying throughout the constraint by –1. For example,

$$h_1(x) > 0$$

is equivalent to

$$-h_1(x) < 0$$

The reverse procedure holds for converting less than inequalities to greater than inequalities.

11.4.4.4 Equalities

Constraints expressed as equalities may be replaced by two inequalities. For example, the constraint

$$h_1(x) = 0$$

is equivalent to

$$h_1(x) \leq 0 \quad \text{and} \quad h_1(x) \geq 0$$

or

$$h_1(x) \leq 0 \quad \text{and} \quad -h_1(x) \leq 0$$

11.4.4.5 Absolute Values

Constraints expressed as absolute values can be replaced by two inequalities. For example,

$$|h_1(x)| \leq \text{constant}$$

is equivalent to

$$h_1(x) \geq -\text{constant} \quad \text{and} \quad h_1(x) \leq \text{constant}$$

and

$$|h_1(x)| \geq \text{constant}$$

is equivalent to

$$h_1(x) \geq \text{constant} \quad \text{and} \quad h_1(x) \leq -\text{constant}$$

11.4.4.6 Inequalities

Inequality constraints can be converted to equality constraints by the introduction of *slack* or *surplus variables*. For example, the constraint

$$h_1(x_1, x_2, ..., x_n) \leq 0$$

may be written as

$$h_1(x_1, x_2, ..., x_n) + x_{n+1} = 0$$

or

$$h'_1(x_1, x_2, ..., x_{n+1}) = 0$$

where x_{n+1} is a slack or surplus variable and represents the amount by which the left-hand side of the original constraint differs from the right-hand side.

Slack and surplus variables are generally assumed to be nonnegative.

11.4.5 Algorithms

The general nonlinear programming forms—(SF1) and (SF2)—use special algorithms in order to get answers. This is discussed, along with applications, later. Designers, planners, ... frequently do the calculations numerically rather than by appealing to special algorithms. Examples of such an approach are given earlier and later.

Depending on the particular optimization technique employed, a constraint may make the calculations easier (for example, generally in linear programming) or harder (for example, generally in calculus) to obtain. The two broad options for handling constraints are as follows:

- Consider the minimization

$$\min J = G(x, u)$$

  directly, unencumbered with constraints, and then examine the constraints to see if they influence this result. In many cases, the unconstrained case is easier to work with than the constrained case.
- Try to incorporate the constraint into the optimization calculations.

For the linear programming form (SF3), algorithms are available that lead to the optimum in a finite number of steps. Computer packages are readily available. Hence, there is always a strong pressure to develop a linear format to optimization in place of nonlinear formats, where the result is much more difficult to come by. Nonlinear programming forms may be dealt with, among other ways, by a series of linearizations and iterations in order to use the good properties of linear programming.

For cases involving discrete variables, it may be possible to deal with them based on the assumption of approximate continuous variables.

11.5 Elementary Optimization

11.5.1 Some Results of Calculus

For the unconstrained case,

$$\min_x J = G(x)$$

calculus gives the conditions for finding extrema (maxima and minima). Briefly, a necessary condition for an extremum is, for G nonlinear and continuous,

$$\frac{\partial G}{\partial x} = 0 \tag{11.1}$$

Values of x satisfying Equation 11.1 are termed *stationary points*. Necessary conditions for a minimum include Equation 11.1 together with

$$\frac{\partial^2 G}{\partial^2 x} \geq 0 \tag{11.2}$$

Such minima are *local* or *relative* minima. All local minima have to be examined in order to obtain the *global* or *absolute* minimum (Figure 11.10).

11.5.2 Equality Constraints

Equality constraints, as given within the form,

$$\min_x J = G(x)$$

$$\text{subject to} \quad h_j(x) = 0 \quad j = 1, \ldots, m < n$$

may be handled in either of two ways. Firstly, the m constraints may be used to eliminate m variables, leaving an unconstrained form in n − m variables. The earlier necessary conditions, Equations 11.1 and 11.2, may then be used. Alternatively, the technique of *Lagrange multipliers* may be used.

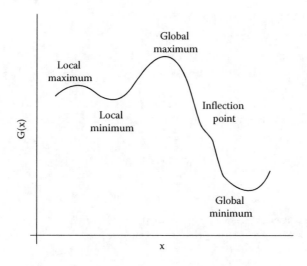

FIGURE 11.10
Extrema.

11.5.3 Lagrange Multipliers

The technique of Lagrange multipliers works by forming a Lagrangian L from the equality constraints adjoined to the objective function through m Lagrange "undetermined multipliers"

$$L(x, \lambda) \triangleq G(x) + \sum_{j=1}^{m} \lambda_j h_j(x)$$

Necessary conditions for a stationary value are

$$\frac{\partial L}{\partial x_i} = 0 \qquad i = 1, \dots, n$$

$$\frac{\partial L}{\partial \lambda_i} (= h_j) = 0 \qquad j = 1, \dots, m$$

(11.3)

These are m + n equations in m + n unknowns, $\lambda = (\lambda_1, \lambda_2, \dots, \lambda_m)^T$ and $x = (x_1, x_2, \dots, x_n)^T$.

11.5.4 Inequality Constraints

Inequality constraints have to be handled numerically via one of the many nonlinear programming algorithms. Alternatively, the form may firstly be looked at without inequality constraints and then the result so obtained checked to see where it lies relative to the inequality constraints and the inadmissible region they define.

Example

The shovel-truck optimization example of earlier may be dealt with using calculus in addition to the already given graphical and numerical approaches.

The two variables, P and K, in the objective function may be reduced to one variable by substituting P from the system equation.

$$J = \frac{100 + 100K}{P} = \frac{100 + 100K}{24\left[K - \dfrac{K^3}{160}\right]}$$

For the unconstrained case, the minimum of J with respect to K could be obtained by examining the stationary values of J. In particular (and dropping the constant 100/24),

$$\frac{dJ}{dK} = \frac{-(1+K)\left[1 - \dfrac{3K^2}{160}\right]}{\left[K - \dfrac{K^3}{160}\right]^2} + \frac{1}{\left[K - \dfrac{K^3}{160}\right]}$$

Setting dJ/dK equal to zero gives

$$2K^3 + 3K^2 - 160 = 0$$

This equation has a root at about K = 3.9, implying that a stationary value for J exists at K = 3.9. The other roots of this equation are complex numbers.

On examination, the second derivative of J with respect to K is positive at K = 3.9 implying a minimum point. The curve of J versus K looks like Figure 11.4. Because only integer numbers of trucks are allowed, the optimum value would be taken as K = 4.

Examine now the effect of including the time-production constraint

$$P \geq 103.3$$

That is,

$$24\left[K - \frac{K^3}{160}\right] \geq 103.3$$

or

$$160K - K^3 - 689 \geq 0$$

Examining the equality part

$$160K - K^3 - 689 = 0$$

This equation has a root between $K = 5$ and $K = 6$. That is, values of K less than 6 do not satisfy the inequality constraint. They are inadmissible values. Hence, the constrained optimal result is $K = 6$, compared with the unconstrained optimal result of $K = 4$. The optimal value lies on the constraint in this case, but this need not always be the case.

Graphically, the formulation in the P–K plane looks like Figures 11.5 and 11.6.

Example

Consider the selection of the proportions of a cylindrical vessel of maximum volume but with a constraint that the surface area be a known value (constant). Introduce the following notation:

J   volume of the vessel
H   height of the cylinder
R   radius of the cylinder
A   constant: surface area (side and ends) of the cylinder

The optimization formulation becomes

$$\max J = \min (-J) = \min -\pi R^2 H$$

$$\text{subject to} \quad 2\pi R^2 + 2\pi R H = A$$

The variables are R and H, which are to be selected so as to maximize the volume.

The calculations may be approached in two ways. Firstly, the constraint may be used to eliminate one of the variables to give unconstrained optimization. Secondly, the idea of Lagrange multipliers may be used to append the constraint to the objective function, again leading to unconstrained optimization, but one which now has an extra variable (the Lagrange multiplier).

For the first method, select H to eliminate. That is,

$$H = \frac{\left(A - 2\pi R^2\right)}{2\pi R}$$

The objective function becomes

$$-J = -\frac{\pi R^2}{2\pi R}\left(A - 2\pi R^2\right) = -\frac{AR}{2} + \pi R^3$$

This has a stationary value given by

$$\frac{d(-J)}{dR} = 0 = -\frac{A}{2} + 3\pi R^2 \quad \text{or} \quad R = \sqrt{\frac{A}{6\pi}}$$

Back-substituting,

$$H = \sqrt{\frac{2A}{3\pi}}$$

Examining the second derivative of −J,

$$\frac{d^2(-J)}{dR^2} = 6\pi R$$

which is always positive. Hence, the above R and H values lead to a (minimum −J) maximum volume result.

The alternative approach is to use a Lagrange multiplier. The Lagrangian becomes

$$L = -\pi R^2 H + \lambda(2\pi R^2 + 2\pi RH - A)$$

and for a stationary value,

$$\frac{\partial L}{\partial R} = 0 = -2\pi RH + \lambda\left(4\pi R + 2\pi H\right)$$

$$\frac{\partial L}{\partial H} = 0 = -\pi R^2 + \lambda\left(2\pi R\right)$$

$$\frac{\partial L}{\partial \lambda} = 0 = -2\pi R^2 + 2\pi RH - A$$

These are three equations in three unknowns R, H, and λ. The last equation is the original constraint. These equations give

$$\lambda = -\sqrt{\frac{A}{24\pi}}$$

and the same values for R and H as before.

Exercise 11.3

For a closed cylindrical tank of given volume, the cost of a side is twice as much as the cost of the top or the bottom. Calculate the least cost proportions of such a cylindrical tank.

Exercise 11.4

For a closed cylinder of given volume, what is the best (in a least cost sense) diameter-to-height ratio, where the top and bottom surfaces cost twice as much per square meter as the side?

Example: Lot Size Forms

In many applications, the objective function that is to be minimized is the sum of two components. One component is an increasing term, while the other component is a decreasing term, for example, Figure 11.11. This occurs, for example, in inventories where the increasing term is of the form bx and the decreasing term is of the form a/x where a and b are constants. The expression "lot size" in the title derives from the early work done in establishing optimum production lots.

The total cost is

$$J = \frac{a}{x} + bx$$

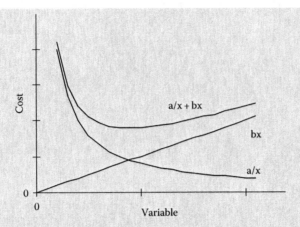

FIGURE 11.11
Lot size characteristics: variable x.

Setting the derivative of J with respect to x to zero gives the optimal lot size as

$$x = \sqrt{\frac{a}{b}}$$

and the corresponding value of the objective function as

$$J = 2\sqrt{ab}$$

Example: Inventory

Inventories imply the stocking of some item, material, etc., with the intent of satisfying a future demand. Inventory decisions involve the determination of how much and when to order (or produce, manufacture, construct, ...) stock. Overstocking implies capital is tied up; there are few periods of shortages and fewer orders are placed (or fewer production runs, ...). Understocking implies less capital is tied up, but there are more occasions when shortages occur and there are more times that orders are placed (or production runs are made, ...). Figure 11.12 illustrates the extreme possibilities. In between these two extremes, there is some optimum (least cost) value as a trade-off. An objective function of cost is implied (Figure 11.13).

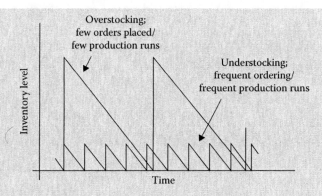

FIGURE 11.12
Overstocking and understocking.

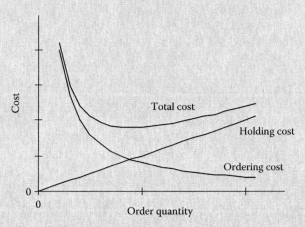

FIGURE 11.13
Cost versus order quantity.

Example: Retail Stock

In one retail outlet, a computer stock inventory package keeps track of the types and amounts of stock that are held. In addition, the package also keeps track of sales and the selling price, buying price, and margins for each of the products. The company aims to stock as little as possible and turn this over as frequently as possible. For example, the company finds it more efficient to keep $100,000 in stock and turn over this stock 10 times a year, rather than keep $1 million in stock and turn over this amount once. The fast turnaround for stock is a principle used in many businesses and especially in the "rag trade." The advantages are that less capital is tied up in stock at any one time, and this reduces storage costs and capital requirements.

If a product cannot be sold within a certain period, then it is not held. The exception to this is stock held in the interest of customer relations.

Example

The literature may use generally the term "inventory control," but this term is avoided here. Also, the inventory discussion here is written in terms of ordering materials, goods, and supplies, but applies equally to in-house production, construction, or manufacturing.

Consider the most elementary (deterministic) inventory case, namely, that involving a *single item* where the *demand* is *constant* over time; there is *instantaneous replenishment* and *no shortages*. Many industrial supplies, hardware supplies, and commonly used materials may fit such a scenario. Figure 11.14 illustrates the change in stock level with time and may be regarded as the model of this inventory situation.

In the notation used in Figure 11.14,

β demand rate

y   highest inventory level

t_0 time between placing an order and reaching zero inventory level again; $t_0 = y/\beta$

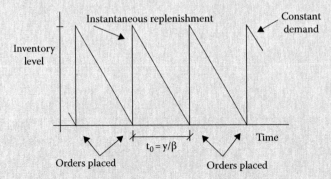

FIGURE 11.14
Single-item, static inventory case.

The overstocking and understocking cases may be seen in Figure 11.12. Some intermediate (minimum cost) value is sought that balances the cost of placing orders and the cost of holding inventory in stock.

The objective function (to be minimized) of cost is made up of two components as follows:

$$\text{Total cost} = \text{Setup cost} + \text{Holding cost}$$

Setup Cost

Setup cost is a single cost associated with placing an order or setting up a production run and may be assumed to be independent of the amount of stock ordered or produced.

Holding Costs

Holding costs reflect the costs of having stock on hand and are assumed to vary with the amount of stock held. They reflect capital invested in the stock, storage and handling charges, depreciation, insurance, and theft.

Introduce the notation:

$C(y)$ total cost per time
K setup cost each time an order is placed
h holding cost per time per item unit

The objective function is

$$C(y)t_0 = K + \frac{hyt_0}{2}$$

The last term, the holding cost per time per item unit, is equal to the area under one of the triangles in Figure 11.14. The whole expression is based on an inventory cycle of $t_0 = y/\beta$. That is,

$$C(y) = \frac{K\beta}{y} + \frac{hy}{2}$$

See Figure 11.15.
For a stationary value,

$$\frac{dC(y)}{dy} = -\frac{K\beta}{y^2} + \frac{h}{2} = 0$$

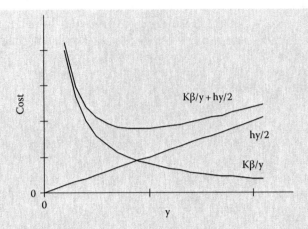

FIGURE 11.15
Inventory: instantaneous replenishment.

or

$$y = \sqrt{\frac{2K\beta}{h}}$$

This is sometimes referred to as Wilson's economic lot size formula.
Examining the second derivative,

$$\frac{d^2C(y)}{dy^2} = \frac{2K\beta}{y^3} > 0$$

This implies a minimum point.
The time between orders, t_0, follows

$$t_0 = \frac{y}{\beta} = \sqrt{\frac{2K}{h\beta}}$$

and the minimum cost

$$C(y) = \sqrt{2K\beta h}$$

Numerous extensions to the earlier formulation and derivation are possible. For example, the deterministic assumptions can be relaxed, and shortages, variable demand, uniform replenishment, buffer stock, delivery lags, price breaks, etc., can be allowed.

Example

A project has the need for bagged material at an average constant rate of (β) 10 bags per day. The cost of placing an order for the material involves a clerk's time and paperwork and is estimated at $50 (K). Holding costs allowing for shed hire, theft, and some spoiling of material is estimated at $1 per bag per day (h). It is assumed that any order is filled instantaneously.

The optimum order level may be calculated using the lot size formula

$$y = \sqrt{\frac{2K\beta}{h}} = \sqrt{\frac{(2)(50)(10)}{1}} = 32 \text{ bags}$$

If it is assumed that 2 days elapse after the placing of the order before the material arrives, then the reorder level becomes 20 bags. That is, when there are 20 bags left in store, place a new order. This value is increased if a buffer stock is used.

The time between orders

$$t_0 = \frac{y}{\beta} = \frac{32}{10} = 3.2 \text{ days}$$

That is, orders may be placed every 3 or 4 days for 30 or 40 bags.

Exercise 11.5

The use of the lot size formula involves being able to estimate ordering costs (personnel, time, stationary, telephone, ...) and holding costs (storage space, personnel, insurance, waste, theft, ...). How would you go about estimating these costs and how good would your estimates be?

Exercise 11.6

Consider a material used frequently in your work. How would you calculate a setup cost and a holding cost?

How would you estimate demand? How realistic is the assumption of determinism for demand?

11.6 Linear Optimization

Many applications have optimization components that are linear in their variables, while in other applications, it is attractive to convert nonlinearities into a linear or a series of linear forms as a way to a result. *Linearity*, here, implies that the contribution of each variable within the objective function and constraints is *proportional* to the variables, and individual contributions are *additive*.

Such linear cases are referred to as *linear programming* (LP) forms and take the standard form (SF3). The component form of SF3 may be written as

$$\min J = \sum_{i=1}^{n} c_i x_i$$

$$\text{subject to} \quad \sum_{i=1}^{n} a_j x_i - b_j \leq 0 \quad j = 1, 2, \ldots, m$$

$$x_i \geq 0 \quad\quad\quad\quad i = 1, 2, \ldots, n$$

or

$$\min J = c_1 x_1 + c_2 x_2 + \cdots + c_n x_n$$

subject to

$$a_{11} x_1 + a_{12} x_2 + \cdots + a_{1n} x_n - b_1 \leq 0$$

$$\vdots$$

$$a_{m1} x_1 + a_{m2} x_2 + \cdots + a_{mn} x_n - b_m \leq 0$$

where a_{ji}, b_j, and c_i are constants.

In the $n = 2$ case, the components, when plotted, may look like Figure 11.16.

The range of applications that can be put into a linear programming form is very broad and includes applications in production scheduling, blending, transportation, assignment, budgeting, planning, bidding strategies, product mix, allocation, and project selection.

11.6.1 Example Linear Cases

Consider some examples of applications that may be made to fit the linear programming form.

11.6.1.1 Production Planning

Applications involving the processing of products or items through different operations may take on linear forms. Typically, the data are given such as

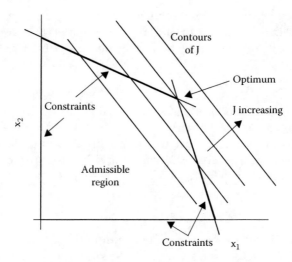

FIGURE 11.16
Example graphical result of an LP form in two dimensions (max J case; or case with negative c_j).

in Table 11.2. Entries in the main body of the table refer to the time (hours, minutes, ...) each product unit takes to pass through each operation. The right-hand column constrains the period (day, week, ...) throughput in each operation. The variables are typically chosen as the number of units of products 1, 2, ..., n made in a period. This gives a linear programming form in n variables and m constraints.

The constraints, being in terms of operation capacities, are written as ≤ inequalities; only when each operation is working to capacity will these inequalities reduce to equalities. The objective function of total output (throughput) or profit is written as a linear combination of the variables; for profit as the objective, the variables are weighted with the product's unit profits. The assumption behind this is that output and profit are proportional

TABLE 11.2

Production Planning Data

Operation	Time/Unit			Operation Capacity (Time/Period)
	Product 1	...	Product n	
1	—		—	—
2	—	...	—	—
.	.		.	.
.	.		.	.
.	.		.	.
m	—		—	—
Profit/unit	—	...	—	

to the quantity produced. Where this is not the case, the objective function will be nonlinear, and the form can only be converted back to a linear programming one through linearization of the objective function.

11.6.1.2 Product Mix

There is a group of applications, where a number of different resources (manpower, materials, machinery, ...) have to combine to give a number of different products, which reduce to a linear programming form. With cost or profit as the objective function, the proportion of resource i, $i = 1, 2, ..., m$, that goes to make up product j, $j = 1, 2, ..., n$ is sought.

11.6.1.3 Transportation

Transportation involves moving some commodity from m origins ($i = 1, 2, ..., m$) to n destinations ($j = 1, 2, ..., n$). There is a cost associated with moving the commodity. There are (supply or capacity) constraints associated with providing the commodity at each origin and (demand or requirement) constraints associated with the commodity at each destination. Each path (i, j) between an origin i and destination j has a cost c_{ij} attached to it. Typical data for such transportation applications follow the form of Table 11.3. The data in the main body of the table are the costs C_{ij}. The right-hand column gives the capacities of the different origins. The bottom line gives the requirements of the different destinations.

The variables are the quantity of commodity carried between origins $i = 1, 2, ..., m$ and destinations $j = 1, 2, ..., n$. That is, for balanced supply and demand, there are $m \times n$ variables. Should a particular route be undesirable or not possible, the cost coefficient associated with that route may be set as a large value. There are m origin capacity constraints and n destination requirement constraints for balanced supply and demand.

TABLE 11.3

Transportation Data

Origin	Destination				Origin Capacity
	1	2	...	n	
1	c_{11}	c_{12}		c_{1n}	—
2	c_{21}	c_{22}	...	c_{2n}	—
.	.			.	.
.	.			.	.
.	.			.	.
m	c_{m1}	c_{m2}		c_{mn}	—
Destination requirements	—	—		—	

11.6.1.4 Assignment

Assignment involves placing a given number of items (people, machinery, ...) in a given number of locations (facilities, situations, ...). The people, machinery, etc., are assigned to locations, situations, etc., in order to maximize the overall effectiveness of the placements.

11.6.1.5 Production Scheduling

Linear programming may be used to schedule production for certain operations. Data for such cases may take the form of Table 11.4.

Not allowing for back orders, Figure 11.17 indicates the relationship between production (supply) in any period and requirements (demand) in any period. The same diagram is repeated for each product.

Overtime can be included and doubles the number of variables.

TABLE 11.4

Production Scheduling Data

Period	Required Production			Available Time
	Product 1	...	Product n	
1	—		—	—
2	—		—	—
.	.		.	.
.	.		.	.
.	.		.	.
m	—		—	—
Time/unit	—		—	

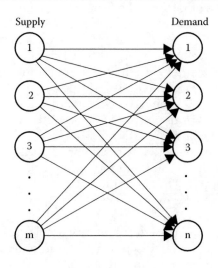

FIGURE 11.17
Supply–demand.

11.6.1.6 Blending

Another group of applications that fits a linear programming form is that where multiple components are mixed together to form either single or multiple end products.

11.6.2 LP Form

Linear programming forms may be dealt with by algorithm (the simplex method), which is coded in many commercially available packages and treated in depth in many texts on optimization, while in two variables, linear programming forms may be dealt with graphically. Such graphical results indicate how the linear programming algorithms work for the many variable case. Consider for example Figure 11.16.

Allowing only positive values for the variables, the variable space is constrained to lie within the positive quadrant. Superimposed on this quadrant are the constraints plotted as equalities. Taking the sense of the constraint inequalities into account defines a region termed the *admissible region* within which the optimum must lie. A point (x_1, x_2) within this region is now sought that maximizes J. To this end, contours of J are drawn on the figure. Clearly, points further from the origin lead to larger J values, and the largest J value is seen to lie at the intersection of the first and third constraints. This intersection point is a *vertex* of the admissible region. The admissible region, since it is made up of a finite number of constraints, has a finite number of vertices.

The fact that the optimum lies at the vertex of the admissible region gives an indication of how an algorithm might work in finding the optimum. A crude algorithm would examine every vertex of the admissible region and choose the vertex that gave the largest value for J. A better algorithm would progress from one vertex to another based on some indicator such as only progressing if the value for J improves. This is the underlying idea of the *simplex method* (developed in the 1940s by Dantzig), which may be found in almost every text dealing with linear programming. The method obtains a global optimum in a finite number of steps. It relies on the admissible region being convex. (The line connecting any two points in the admissible region lies within the admissible region.) Its calculations are intended for a computer and not by hand.

Note that where the objective function has a gradient parallel to a constraint, there arises the possibility of nonunique values; all points on that constraint are optimal. This is of no embarrassment because the non-uniqueness gives the user more flexibility in choosing suitable values for the variables.

This explanation generalizes to the many variable cases.

The dual form, which is not discussed here, can provide additional useful information on the linear programming form.

Exercise 11.7

Use graphical means to determine the optimum for the following LP form:

$$\min J = -2x_1 - x_2$$

$$\text{subject to} \quad x_1 \geq 0 \quad x_2 \geq 0$$

$$-2x_1 + x_2 \leq 4$$

$$x_1 + 3x_2 \geq 3$$

$$x_1 + x_2 \leq 7$$

Exercise 11.8

Use graphical means to determine the optimum for the previous exercise, but with an objective of

$$\min J = -x_1 - x_2$$

Exercise 11.9

A labor-intensive maintenance program is to be carried out over a short period, working 24 h each day. The labor requirement in any 24 h period is as follows:

Time	Minimum Labor Requirement
6:00–10:00	5
10:00–14:00	10
14:00–18:00	6
18:00–22:00	10
22:00–2:00	4
2:00–6:00	2

Each person could be expected to work an 8 h shift in any 24 h period. Formulate, in LP terms, the situation where the minimum number of persons that satisfies the earlier requirements is sought.

Exercise 11.10

On a project, the minimum number of trucks required on day i (i = 1, 2, ..., 25) is b_i. Each truck could be expected to operate for 6 consecutive days before having minor maintenance work carried out. An extra cost c_i per truck day is incurred for every truck working on day i above the minimum b_i requirement. With an objective of (minimizing) cost, formulate this in an LP form.

Exercise 11.11

A company makes two products, product X and product Y, and each product has to go through two operations during manufacture. The second operation may be undertaken in either of two facilities, B or C. The first operation can only be carried out in facility A. The data for this manufacturing case are as follows:

	Product X	Product Y	Maximum Weekly Operating Hours
Operation 1			
Facility A hours/unit	0.4	0.6	72
Operation 2			
Facility B hours/unit	0.2	1.6	72
Operation 2			
Facility C hours/unit	1.2	1.2	72 regular 36 overtime

The hourly costs of facilities A, B, and C are $100, $120, and $130. Facility C may be also operated after regular working hours at an hourly cost of $150. Products X and Y sell for $500 and $480, respectively. A minimum of 20 of product Y must be produced each week to satisfy an existing contract. Calculate the product mix giving optimum profit and mode of operation (including which facilities are to be used and whether regular or overtime working hours are used).

11.6.3 Multiple Objective Linear Programming

Many applications have multiple objectives. For example, a project manager might like a project to be finished in minimum time and with minimum cost.

Multi-objective linear programming (MOLP) involves two basic steps:

1. Determining the set of nondominated values
2. Determining the best compromise value from the set of nondominated values

A graphical method is used to demonstrate the approach.
The general multi-objective form might be posed as

$$\min J = \left[J_1, J_2, \ldots, J_p \right]$$

$$\text{subject to} \quad g_j(x) \le b_j \quad j = 1, 2, \ldots, m$$

$$x_i \ge 0 \quad i = 1, 2, \ldots, n$$

where J_k, $k = 1, \ldots, p$ are functions of the variables x_i, $i = 1, \ldots, n$. J_k and g_j are linear functions of the variables x_i, and b_j ($b_j > 0$; $j = 1, \ldots, m$) are constants.

The ideal result would be to find that admissible set of variables x_i ($i = 1, \ldots, n$) that would minimize the individual objective functions simultaneously. However, with conflicting objectives, an admissible value that optimizes one objective function may not optimize any of the other objective functions. This means that what is optimal in terms of one of the p objectives is, generally, not optimal for the other $p - 1$ objectives. Accordingly, there is a need for a new interpretation as to what constitutes an optimal result in the multiple objective case.

Example

To illustrate graphically, consider the following two-variable, two-objective function example:

$$\max J = \left[J_1, J_2 \right]$$

$$\text{where} \quad J_1 = 5x_1 + x_2$$

$$J_2 = x_1 + 4x_2$$

$$\text{subject to} \quad x_1 + x_2 \leq 6$$

$$x_1 \leq 5$$

$$x_2 \leq 3$$

$$x_1, x_2 \geq 0$$

For two variables and two objective functions, the set of nondominated values can be shown readily by graphical means. To illustrate this, call the space defined by the x_1 and x_2 axes the variable space. For the purpose of multi-objective programming, the space that is defined by the J_1 and J_2 axes is called the objective space, to distinguish it from the variable space. Any extreme point of the variable space can be transferred to the objective space with J_1 and J_2 values (objective function values) calculated at that extreme point of the admissible space. Consider the earlier example. The admissible region and extreme points are shown in the variable space of Figure 11.18a. In Figure 11.18b, the same admissible region and its extreme points are shown in the objective space.

It is seen from Figure 11.18b that the extreme points C and D are nondominated. Consequently, all points defined by J_1 and J_2 values on the line segment connecting the two adjacent extreme points C and D in the objective space are also nondominated. This means that the points of the line segment CD (including the extreme points C and D), which is darkened in Figure 11.18b, constitute the complete set of nondominated objective values. Accordingly, all points defined by x_1 and x_2 on the line segment CD (including C and D) in the variable space, darkened in Figure 11.18a, constitute the complete set of nondominated values. Thus, both the variables set and the objective set have been graphically determined.

The following general rule, called the north-east rule, locates graphically the nondominated set of objective values in the objective space: *When all objectives are to be maximized, an admissible value is nondominated, if there are no admissible values lying to the north east.*

Once the set of nondominated values has been found, the user has to articulate his/her preferences and establish a system of comparative evaluations between levels of achievement of the conflicting objectives. For this reason, methods of MOLP, other than the graphical method, are interactive. In order to arrive at a single value, the user has to trade off nondominated values against each other.

Methods are available, which aid the user in making realistic judgments in selecting the best compromise value. Such methods include weighting, sequential, step, and constraint methods.

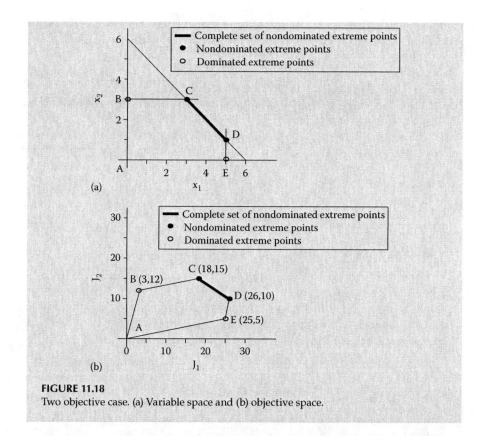

FIGURE 11.18
Two objective case. (a) Variable space and (b) objective space.

Exercise 11.12

What is the underlying reasoning behind the north-east rule?

11.7 Nonlinear Programming

Many applications fit within the domain of nonlinear programming, which is a collection of techniques and algorithms for forms SF1 and SF2. To fully understand the complete picture of nonlinear programming requires a solid foundation in linear algebra and calculus. Numerous techniques have been proposed. However, there does not appear to be one technique that is suitable for all possible cases. All techniques are based on numerical algorithms.

Some of the more common techniques are presented later, but it is by no means a complete presentation. The presentation is also given without complete mathematical rigor. The development proceeds from the unconstrained case, and from single variable (univariate) to multiple variable (multivariate) cases.

For the single-variable case, the objective function is one dimensional; it may be unimodal or have a number of turning points whereupon the algorithm may converge to a local optimum rather than the global optimum. A similar situation is repeated in the multivariable case where the objective function is multidimensional. Different starting values may be tried in an attempt to converge to different local optima and the global optimum. Different techniques demonstrate different convergence properties, which are also dependent on the shape of the objective function and the constraints.

Firstly consider an example nonlinear case.

Example

Multiple items in an inventory may be treated in a similar fashion to the single-item case discussed earlier. Consider the special case where different items are competing for the same storage space.

Let there be n items competing for the limited storage space A, with each item requiring storage space a_i per item unit. That is,

$$\sum_{i=1}^{n} a_i y_i \leq A$$

Using instantaneous replenishment and no shortage assumptions, and denoting variables with a subscript i when referring to item i, the total cost becomes

$$C(y_1, y_2, \ldots, y_n) = \sum_{i=1}^{n} \left[\frac{K_i \beta_i}{y_i} + \frac{h_i y_i}{2} \right]$$

The task of selecting the optimal order levels for the n items is one of n variables $y_1, y_2, \ldots, y_n$, and one involving the minimization of C subject to the inequality constraint on storage.

The result can be reasoned as follows. If the constraint is inactive, that is the optimal value does not lie on the constraint, then an unconstrained result may be obtained as for the single-item case, and

$$y_i = \sqrt{\frac{2K_i \beta_i}{h_i}}$$

If the constraint is active, that is, the optimal value lies on the constraint, then the constraint becomes an equality and may be appended to the cost using a Lagrange multiplier, λ, to form a Lagrangian:

$$L(y_1, y_2, \ldots, y_n, \lambda) = C(y_1, y_2, \ldots, y_n) + \lambda\left[A - \sum_{i=1}^{n} a_i y_i\right]$$

$$= \sum_{i=1}^{n}\left[\frac{K_i \beta_i}{y_i} + \frac{h_i y_i}{2}\right] + \lambda\left[A - \sum_{i=1}^{n} a_i y_i\right]$$

For a stationary value,

$$\frac{\partial L}{\partial y_i} = 0 = -\frac{K_i \beta_i}{y_i^2} + \frac{h_i}{2} - \lambda a_i \quad i = 1, 2, \ldots, n$$

$$\frac{\partial L}{\partial \lambda} = 0 = A - \sum_{i=1}^{n} a_i y_i$$

From these $n + 1$ equations, the $n + 1$ optimal values for $y_1, y_2, \ldots, y_n$ and λ can be found.

A comparison of the constrained and unconstrained results in any particular case will establish whether the constraint is active or not.

Consider some numerical values. A project site storage shed is used to stock bags of chemicals AA, BB, and CC. Respectively for each, the costs of placing orders are $50, $60, and $80, the daily demands are 10, 2, and 5 bags per day, while the holding costs are estimated at $1, $0.50, and $2 per bag per day. Each bag occupies about 0.05 m³, and the total storage space available is 6 m³.

The data may be summarized as follows:

Item	i	K_i ($)	h_i ($/Bag/Day)	a_i (m³)	β_i (Bags/Day)
Chemical AA	1	50	1	0.05	10
Chemical BB	2	60	0.5	0.05	2
Chemical CC	3	80	2	0.05	5

The unconstrained result is

$$y_1 = \sqrt{\frac{(2)(50)(10)}{1}} = 32 \text{ bags}, \quad t_{01} = \frac{32}{10} = 3.2 \text{ days}$$

$$y_2 = \sqrt{\frac{(2)(60)(2)}{0.5}} = 22 \text{ bags}, \quad t_{02} = \frac{22}{2} = 11 \text{ days}$$

$$y_3 = \sqrt{\frac{(2)(80)(5)}{2}} = 20 \text{ bags}, \quad t_{03} = \frac{20}{5} = 4 \text{ days}$$

Checking the value of the constraint,

$$\sum_{i=1}^{n} a_i y_i = 0.05(32 + 22 + 20) = 3.7 \text{ m}^3 < 6 \text{ m}^3$$

That is, the constraint is not violated, and the unconstrained result is the optimal result.

Assume, for demonstration purposes, that the available storage space A is 3 m³. The constrained result may be obtained iteratively by guessing values for λ and checking the left- and right-hand sides of the storage constraint. Some typical calculations are as follows. The y_i are calculated from the first n stationary value conditions, namely,

$$y_i = \sqrt{\frac{2K_i \beta_i}{h_i - 2\lambda a_i}} \quad i = 1, 2, \dots, n$$

λ	y_1 (Bags)	y_2 (Bags)	y_3 (Bags)	$0.05(y_1+y_2+y_3)-3$
0	31.6	21.9	20.0	0.677
−2	28.9	18.5	19.1	0.322
−4	26.7	16.3	18.3	0.066
−4.2	26.5	16.2	18.2	0.044
−4.4	26.4	16.0	18.1	0.022
−4.6	26.2	15.8	18.0	0.001
−4.8	26.0	15.6	18.0	−0.020
−5	25.8	15.5	17.9	−0.040
−5.2	25.6	15.3	17.8	−0.060

A λ value of −4.6 and y_1, y_2, and y_3 values of 26, 16, and 18 bags give a result that is minimal in cost and lies on the storage constraint.

Exercise 11.13

A particular process utilizes four components more heavily than other items. For calculation purposes, assume constant demand rates, no shortages, and instantaneous replenishment upon placing orders. The data are as follows:

Component i	K_i ($)	h_i ($/cpt/Day)	a_i (m²)	β_i (cpt/Day)
1	40	0.25	1	25
2	10	0.15	1	15
3	90	0.15	1	15
4	80	0.25	1	10

For a total storage space availability of 100 m², calculate the optimal ordering level for each item.

11.7.1 Unconstrained Case

11.7.1.1 Single Variable

Methods that deal with single variables might be classified as

- Interval elimination methods
- Function approximation methods
- Hybrid methods

11.7.1.2 Interval Elimination Methods

Interval elimination methods work by repeatedly making the interval smaller within which the optimum x value lies. There are a number of search methods that fall within this category. All differ in the way the interval size, within which the optimum x lies, is reduced from one iteration to the next.

The assumption of unimodality of the function $G(x)$ is required for the elimination techniques to work. A unimodal function is one with a single hump or a single valley.

Superscripts on x in the following refer to particular values of x.

Example

Consider a function on the interval [0,1].

Calculating the values of the function at x^1 and x^2 (values of x) gives the diagram, for example, of Figure 11.19a. From the unimodality assumption, the minimum cannot lie in the interval $[x^2, 1]$, and this interval can be discarded.

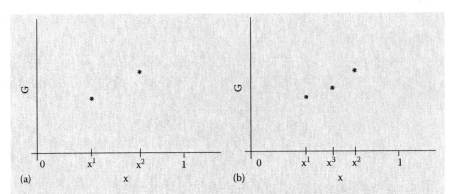

FIGURE 11.19
Example search method. Numbers after the variable indicate values of that variable. (a) First two evaluations and (b) first three evaluations.

Calculating the value of the function at x^3 gives the diagram, for example, of Figure 11.19b. Clearly, again based on the assumption of unimodality, the minimum cannot lie in the interval $[x^3, x^2]$, and this interval may be discarded.

The process is repeated, each time further reducing the interval size within which the optimum lies.

The various proposed search methods differ in the algorithms they use for selecting $x^1, x^2, x^3,\ldots$

Exercise 11.14

An elementary, and perhaps not too efficient, way of selecting x^1, $x^2, x^3, \ldots$ is to select x^i that halves the remaining interval.

Given an objective function

$$J = \frac{1+K}{K - \left(K^3/160\right)}$$

calculate J at $K = 1$ and $K = 7$. Now approximately halve the interval $[0,7]$, and eliminate that half in which the minimum does not lie.

Repeat the process. How many steps does it take to reach the optimum?

11.7.1.3 Function Approximation Methods

Function approximation methods replace the original objective function by a simple form, such as a quadratic function, at each iteration. The optimum of the replacing function is found using calculus, and a new iteration is repeated. Included in the function approximation methods is *Newton's method*, also called the *Newton–Raphson method*.

Newton's method uses the first three terms of a Taylor's series expansion of G(x) about the current iteration point, x^k

$$g(x) = G(x^k) + G'(x^k)(x - x^k) + \frac{1}{2}G''(x^k)(x - x^k)^2$$

to give

$$x^{k+1} = x^k - \frac{G'(x^k)}{G''(x^k)}$$

as the result for G'(x) = 0. This is Newton–Raphson applied to finding the root of an equation where the equation is G'(x) = 0 (Figure 11.20).

Here G' and G" denote the first and second derivatives of G with respect to x. g(x) is a quadratic approximation to G(x)

Figure 11.21 shows the process.

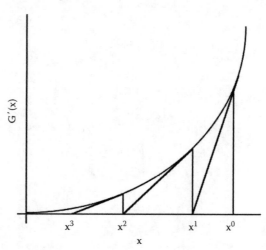

FIGURE 11.20
Newton–Raphson applied to finding the root of an equation.

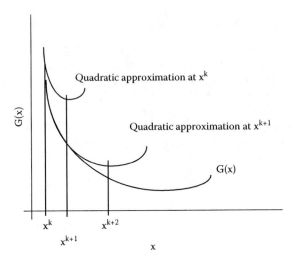

FIGURE 11.21
Newton's method.

Example

For the exercise of finding the proportions of a cylinder of maximum volume considered earlier,

$$J = \pi R^3 - \frac{AR}{2}$$

$$\frac{dJ}{dR} = 3\pi R^2 - \frac{A}{2}$$

$$\frac{d^2J}{dR^2} = 6\pi R$$

Then

$$R^{k+1} = R^k - \frac{\left[3\pi R^{k2} - (A/2)\right]}{6\pi R^k}$$

Exercise 11.15

For this example, start with $R^0 = 0.5$ and carry out several iterations to find the optimal radius. Assume $A = 6\pi$. The exact answer is $R = 1$.

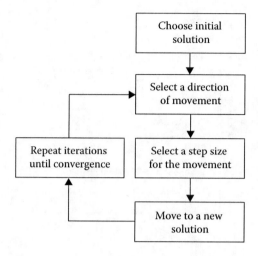

FIGURE 11.22
General nonlinear programming approach.

11.7.1.4 Multiple Variables

General approaches to the multivariable case adopt a process as illustrated in Figure 11.22. The selection of the direction of movement and the step size distinguish many of the techniques.

Most nonlinear programming methods are algorithmic and get to the minimum point through a series of iterations from a guessed starting point. The art is in setting the iteration step size—too small and the calculations end up with too many iterations; too big and the calculations may overshoot the optimum point.

The methods may only home in on local optima, and so several starting points may be tried in an attempt to find the global optimum.

11.7.1.5 Gradient Method

The gradient method gives the direction of movement as that of the tangent or gradient at the current value (Figure 11.23). The technique is known as the method of steepest descent or steepest ascent depending on whether minimization or maximization, respectively, is involved.

The gradient method starts at one point on the curve and goes in the direction of the gradient to the bottom of the curve in a series of finite-sized steps. When the gradient equals zero (or small), the iterations stop, and this point is assumed to be the minimum.

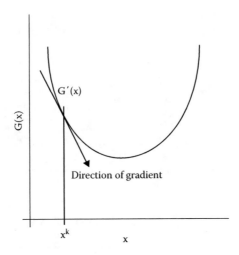

FIGURE 11.23
Gradient method.

Exercise 11.16

For the earlier exercise of finding the maximum volume of a cylinder, plot contours of the objective function in the plane having axes of H and R. Work the gradient method on these contours.

For the earlier shovel-truck optimization, plot contours of the objective function in the plane having axes of P and K. Work the gradient method on these contours.

11.7.1.6 Newton's Method

The treatment of Newton's method for a single variable extends to the multivariable case. The method gives both the direction of movement and the step size.

Some variations on this method, such as in the *method of conjugate directions* and the *variable metric methods*, have been proposed.

11.7.2 Constrained Case

Numerous methods have been proposed for general (constrained) nonlinear programming and include methods based on the Kuhn–Tucker conditions (a generalization of the Lagrange multiplier approach), gradients, and penalty functions.

11.7.2.1 Kuhn–Tucker Conditions

Using Lagrange multipliers to adjoin equality constraints to the objective function, necessary conditions for an optimum can be obtained as reported earlier. The extension of these conditions to the case with inequality constraints is the Kuhn–Tucker conditions.

The Lagrangian is

$$L(x,\lambda) \triangleq G(x) + \sum_{j-1}^{m} \lambda_j h_j(x)$$

assuming that h_j represents all the constraints, including any nonnegativity constraints. λ_j is a generalized Lagrange multiplier.

The necessary conditions for an optimum in essence are

$$\frac{\partial L}{\partial x} = 0 \quad \text{where} \quad \lambda_j \begin{cases} \geq 0 & \text{for } h_j(x) = 0 \\ = 0 & \text{for } h_j(x) < 0 \end{cases}$$

That is, where the optimum lies on the constraint boundary $\lambda_j \geq 0$, and where the optimum lies within the admissible region $\lambda_j = 0$.

The Kuhn–Tucker conditions are used to test a point for relative minimum properties or their basis is used to find a point that decreases the value of the objective function.

11.7.2.2 Quadratic Programming

Quadratic programming involves a quadratic objective function and linear constraints:

$$\min J = \sum_{i=1}^{n} \left(c_i x_i + \sum_{k=1}^{n} d_{ki} x_k x_i \right)$$

$$\text{subject to} \quad \sum_{i=1}^{n} a_{ij} x_i - b_j \leq 0 \quad j = 1, 2, \ldots, m$$

$$x_i \geq 0 \qquad\qquad i = 1, 2, \ldots, n$$

For such a formulation, the Kuhn–Tucker conditions could be expected to be both necessary and sufficient conditions for optimality.

There are a number of suggested algorithms for quadratic programming.

11.7.2.3 Penalty Methods

Penalty methods work by converting a constrained formulation into an unconstrained one. The unconstrained case may then be dealt with by a number of suitable techniques such as the gradient method.

For each constraint

$$h_j(x) \le 0 \quad j = 1, 2, \ldots, m$$

a penalty term P(x) is added to the objective function; it penalizes any value of x that violates this constraint. P(x) has the following properties:

$$P(x) \begin{cases} = 0 \text{ if } x \text{ is a feasible solution} \\ < 0 \text{ if } x \text{ is an infeasible solution} \end{cases}$$

That is,

$$P(x) \begin{cases} = 0 & \text{if } h_j(x) \le 0, \quad \text{all } j \\ < 0 & \text{otherwise} \end{cases}$$

A suitable form for P(x) might be

$$P(x) = -\sum_{j=1}^{m} \left(\max\left[0, h_j(x) \right] \right)^2$$

although this may not, in all cases, penalize inadmissible values enough.

Variations on the penalty methods include the barrier methods, which add a barrier function to the objective function; it penalizes values of x near to the boundary of the admissible region, thereby preventing iterations from moving from admissible to inadmissible points.

11.8 Other Optimization Forms

All the earlier formulations fit a classification, which might be called *static optimization*. The components do not change with time.

Systems that vary with time are usually modeled as differential equations or difference equations (see state equation models). The associated objective functions become summations or integrals over time. The optimization formulation is correspondingly more complicated than the formulations considered earlier.

Such *dynamic optimization* cases can be dealt with via the *calculus of variations* or *Pontryagin's maximum principle*. The latter is like a generalization of the Lagrange multiplier idea to the dynamic case. Both approaches lead to necessary conditions, from which an optimum can be found.

An alternative technique is *Bellman's dynamic programming*, which, in the dynamic version discretized over time, develops the optimum in stages proceeding in increments of time. Dynamic programming can also be applied to static cases that show some form of staging.

Such formulations and approaches are popular in the area of control systems where the general field is known as *optimal control (systems) theory*.

Generalizations are possible to allow a spatial variable to replace the time variable and to optimization over the four-dimensional time–space domain.

A reference covering the earlier topics is Carmichael (1981).

12

Decision Approaches and Tools

12.1 Introduction

Common decision making typically goes through the synthesis via iterative analysis steps:

- Definition
- Objectives and constraints statement
- Alternatives generation
- Analysis and evaluation
- Selection

Feedback and iterative modification, for clarification and refinement purposes, may occur.

Most of what the literature terms "decision theory" or "decision analysis," however, concerns itself with just two of these steps, namely, alternatives generation, and analysis and evaluation. It tends to cover idealized and elementary decision situations under uncertainty, but its study can give insight to, and structure for more complicated and more realistic decision making. A distinguishing feature is that it attempts to deal with uncertainty around decisions; in doing so, the trade-off is that only elementary situations are able to be considered. Real-world decision making is often so complex that a combination of experience, judgment, and theoretical tools is necessary.

12.1.1 Terminology

"Decision theory" is well established and has ventured off into its own unique terminology and notation domain. As such, it uses terms that conflict with usage in the rest of the book. For example,

- "Analysis" in "decision analysis" refers to a dissecting of the decision making, and not the preferred sense of referring to the analysis configuration.

- "Theory" in "decision theory" and "game theory" refers to "formal and quantitative procedures associated with decision making, as opposed to practice"; it is not used in the sense discussed in other chapters.
- "Model" in "decision model" is used in the sense of modeling the decision making, and not in the preferred sense of a system model.
- "Uncertainty" refers to possible outcomes known, but without information on the likelihood of occurrence. While "risk" refers to possible outcomes known, an estimate of each outcome occurring can be made. This is quite different usage to elsewhere in the book.
- "States of nature" refers to possible future events that might occur. The decision maker does not know which event will occur. That depends on chance or nature. These states of nature are not under the influence of the decision maker. A state of nature is not the same as system state used elsewhere in the book; a system state follows from the choice of control/decision.
- "Payoff" refers to the result of a particular decision being made in conjunction with a subsequent state of nature occurring. Payoff is equivalent to the use of the term "objective" adopted everywhere else in this book.

To avoid confusion, the terms "analysis," "theory," "model," "uncertainty," and "risk," in the as used in "decision theory," are not used in this chapter. Rather, if used, they are to be taken in the sense of the rest of the book.

The terms "decision," "control," and "action," when used in the sense of that which is chosen by the engineer, are used interchangeably here. Generally, the term "decision" or "action" is the established term used in the "decision theory" literature.

Outline: The early sections look at a basis for making decisions or choosing between alternatives. Although only elementary situations are examined, the conceptual approach will help in more complicated situations. Section 12.4 alerts you to the added difficulties of including a competing decision maker. The chapter then discusses selecting between alternatives when there is a sequence of decisions to be made. This decision process may be represented graphically by decision trees. Where additional information becomes available, Bayes theorem allows this information to improve the decisions made. The trees may be analyzed through expected value or utility measures.

> I used to be indecisive ... but now I'm not so sure.
>
> **Anon.**

12.2 Underlying Framework

Situations in which decisions are made may be classified according to Table 12.1.

For each situation, approaches have been developed, albeit for only elementary versions of each situation. The framework for decision making of Table 12.2 is used.

The earlier information might be displayed in the form of a payoff matrix.

TABLE 12.1

Decision Situations

One-off	Decisions that are made once and are not repeated
Many-off	Decisions that are repeated many times
Single stage	One isolated decision is made
Multi-stage	A sequence of decisions is made
Certainty in decision making	Determinism. Outcomes are known
Noncertainty in decision making	(i) Possible outcomes are known but without information on the likelihood of occurrence
	(ii) Possible outcomes are known, and an estimate of the probability of each outcome occurring can be made
Competitive decision making (gaming approaches)	Where the decision outcome is affected by the actions of an opponent

TABLE 12.2

Underlying Framework

Decision alternatives	Alternative courses of action that the decision maker has available.
States of nature	A list of possible future events that might occur. The decision maker does not know which event will occur. That depends on chance or nature. These states of nature are not under the influence of the decision maker.
Payoff	The result of a particular decision being made in conjunction with a subsequent state of nature occurring.

Example

For the earthquake design of a small bridge, the payoff may be in terms of damage cost (Table 12.3).

Entries in Table 12.3 refer to the cost of damage should a particular earthquake occur and having chosen a particular level of design.

If there is information about the relative probabilities of the states of nature arising, the information can be represented in a probability matrix (Table 12.4).

Entries in Table 12.4 refer to the probabilities of the states of nature occurring.

In the case of a decision with noncertainty type (i), there is only enough information to create the payoff matrix for the decision. In the case of a decision with noncertainty type (ii), both the payoff matrix and the probability matrix can be constructed.

TABLE 12.3

Example Payoff Matrix (Damage Cost)

Decision Alternative—Magnitude of Earthquake Designed For	States of Nature—Magnitude of Earthquake Incurred		
	0–4	4–7	>7
None	3000	5000	8000
0–4	0	3000	7000
4–7	0	0	4000
>7	0	0	0

TABLE 12.4

Example Probability Matrix

Decision Alternative	States of Nature		
	0–4	4–7	>7
None	0.6	0.3	0.1
0–4	0.6	0.3	0.1
4–7	0.6	0.3	0.1
>7	0.6	0.3	0.1

12.3 Ranking Payoffs

The ranking of payoffs varies depending on the noncertainty type. Noncertainty type (i) is where the probabilities of possible outcomes (states of nature) are not known. Noncertainty type (ii) assumes that a probability can be associated with every possible state of nature.

TABLE 12.5

Example Developer Data

	States of Nature		
	Low Demand for Land	Moderate Demand for Land	High Demand for Land
Develop small subdivision	$3,500,000	$3,000,000	$2,700,000
Develop medium subdivision	$1,000,000	$12,500,000	$12,400,000
Develop large subdivision	−$500,000	−$250,000	$25,000,000

The probability of competitors or others actively trying to influence outcomes is not included. These and other influences are discussed later in relation to games.

12.3.1 Example

To demonstrate the different ways that payoffs are ranked, consider an example. A real-estate developer wants to develop a new subdivision. The payoff table for potential profits is given in Table 12.5.

12.3.1.1 Noncertainty Type (i)

Noncertainty type (i) refers to there being no probabilities associated with the states of nature.

12.3.1.1.1 Maximin

The decision maker selects the action, which has the largest minimum payoff arguing that, no matter what happens, the return will be the maximum of the minima. In this example, minimum payoffs for each alternative decision made are $270,000 (develop small subdivision decision), $1,000,000 (develop medium subdivision decision), and -$500,000 (develop large subdivision decision). The largest of these is $270,000, and hence the maximin decision maker would select the small development alternative. This approach is very conservative or *pessimistic*, being based on extremes.

A disadvantage with maximin is that it disregards a lot of the information in the payoff matrix by looking only at minimum payoffs.

12.3.1.1.2 Maximax

In contrast, maximax is adopted by the very *optimistic* aggressive decision maker or gambler. In this example, the maximum payoffs for each alternative decision made are $3,500,000 (develop small subdivision decision), $12,500,000 (develop medium subdivision decision), and $25,000,000 (develop large subdivision decision). The largest of these is $25,000,000, and hence using maximax would lead to the development of the large subdivision alternative, in the hope that there will be a large demand.

TABLE 12.6

Opportunity Loss

	Low Demand	Moderate Demand	High Demand
Small subdivision	0	$9,500,000	$22,300,000
Medium subdivision	$2,500,000	0	$12,600,000
Large subdivision	$4,000,000	$12,750,000	0

This has similar disadvantages to maximin. By looking at only one possible outcome, it neglects most of the relevant information.

12.3.1.1.3 Minimax Regret

This conservative approach uses an opportunity loss table to determine the maximum opportunity loss (or regret) that could occur from each decision and selects the *minimum* of the *maximum regret* values. The opportunity loss or regret is given from the payoff table by

$$R(a_i, S_j) = |V^*(S_j) - V(a_i, S_j)|$$

$R(a_i, S_j)$ is the opportunity loss for decision a_i and state of nature S_j
$V^*(S_j)$ is the maximum payoff under state of nature S_j
$V(a_i, S_j)$ is the payoff associated with decision a_i and state of nature S_j

Minimax regret has similar issues to maximin, with the added flaw that the ranking of choices depends on the choices specified. Minimax regret is also sensitive to the choices compared.

In this example, the opportunity loss table is given in Table 12.6.

The maxima for each decision are $22,300,000, $12,600,000, and $12,750,000. Of these, $12,600,000, corresponding to the medium subdivision alternative, is the minimum. If minimax regret is applied, then the medium subdivision alternative is selected.

12.3.1.1.4 Summary

Each of the ranking methods (here and others not mentioned) for decision making under noncertainty type (i) has deficiencies. Each in its own way fails to use all the relevant information.

12.3.1.2 Noncertainty Type (ii)

Noncertainty type (ii) refers to there being probabilities associated with the states of nature.

12.3.1.2.1 *Expected Monetary Value*

Expected monetary value (EMV) has popular usage. If probabilities $P(S_1)$, ..., $P(S_n)$ are assigned to the n states of nature, a weighted average payoff for the decision maker's action, a_i, can be calculated:

$$EMV(a_i) = \sum_{j=1}^{n} P(S_j)V(a_i, S_j)$$

Once the expected value for each action is calculated, the decision with the largest expected value is chosen.

For one-off decisions, the expected value technique can be criticized. Over the long term, or over many similar decisions, it is reasonable.

In the earlier real estate example, if the probability of 0.8 is associated with low demand, 0.1 with moderate demand, and 0.1 with high demand, the EMVs would be

$$EMV(a_1) = 0.8(3,500,000) + 0.1(3,000,000) + 0.1(2,700,000)$$

$$= \$3,370,000$$

$$EMV(a_2) = \$3,290,000$$

$$EMV(a_3) = \$2,075,000$$

Therefore, using EMV, the small subdivision alternative is chosen.

12.3.1.2.2 *Expected Opportunity Loss*

Expected opportunity loss (EOL) is related to EMV in a way that parallels the relation between minimax regret and maximin. The expected value of the opportunity loss for each decision alternative is calculated and the minimum is chosen.

In the real estate example,

$$EOL(a_1) = 0.8(0) + 0.1(9,500,000) + 0.1(22,300,000)$$

$$= \$3,180,000$$

$$EOL(a_2) = \$3,260,000$$

$$EOL(a_3) = \$4,475,000$$

Alternative 1 would therefore be chosen.

Exercise 12.1

Suppose that a decision maker, having four decision alternatives and four states of nature, has developed the following payoff table.

Decision Alternative	State of Nature			
	Q_1	Q_2	Q_3	Q_4
a_1	15	13	12	9
a_2	12	11	9	10
a_3	9	10	7	12
a_4	7	10	8	12

Assume that the decision maker knows nothing concerning the probability of occurrence of the various states of nature.

Using maximin, which decision alternative would be selected?

Using maximax, which decision alternative would be selected?

Using minimax regret, which decision alternative would be selected?

Assume that the following estimates of probability of occurrence of the states of nature can be made:

$$P(Q_1)=0.3$$
$$P(Q_2)=0.2$$
$$P(Q_3)=0.2$$
$$P(Q_4)=0.3$$

Which decision would be selected based on EMV?

Exercise 12.2

You have decided to dispose of your car. Unfortunately, it has failed its safety inspection and will require $1000 of repair to obtain an inspection pass. A wrecker has offered you $500 for the car as it is. You have a little idea of what the market is like for your type of car; if, with the car having passed inspection, the market is strong you would get $2500, if moderate $1500, and if weak $800. Without inspection, you could sell the car for $1200, $800, and $500 in strong, moderate, and weak markets, respectively. If you decide to advertise your car for sale, you incur a $100 advertising cost.

Identify the three possible strategies and construct the payoff matrix. What decision would you adopt if

(i) You are a pessimist
(ii) You are an optimist
(iii) You wish to minimize your regret

Suppose the probabilities of a strong, moderate, and weak market are given by 0.3, 0.3, and 0.4, respectively. Which decision maximizes your expected value?

Exercise 12.3

A football club orders programs for its home games in batches of 10,000, 20,000, or 30,000 at costs of $1,000, $1,600, or $2,000, respectively. Programs sell at 20¢ each. The order for each match has to be placed a week in advance, and any program not sold on the day has no value. Sales of program depend on the match attendance. For simplicity, assume that the program sales will be 10,000, 20,000, or 30,000. Find the maximin and minimax regret decisions for program orders.

Next Saturday, the club will be playing its arch rival. The director of the club estimates the probability of 30,000 sales is 0.6, 20,000 sales is 0.3, and 0.1 for 10,000 sales. What is the optimum EMV decision under these conditions?

Exercise 12.4

The government is proposing a major motorway development in an area where there is land available to build a hotel/restaurant. Local protesters are raising objections to the government's intentions, and it is by no means certain that the government will go ahead. If the development is confirmed, the hotel will yield a profit equivalent to a present value of $20 million. On the other hand, if the land is bought, but the development does not go ahead, the present value of the resultant loss is $5 million. There is an alternative site for this investment, which will yield a return of $7 million irrespective of the government's decision on the motorway development.

What is the smallest value of the probability that the development will go ahead, which will cause you to buy the land? How sensitive is your answer to the assumed return of $20 million?

12.4 Decisions with Competition or Conflict

Often the decision maker has to make a decision where other persons may try to influence the outcome of that decision. Many business decisions and engineering decisions fall into this category. For example, in tendering, there will usually be competitors.

In an attempt to deal with these types of decisions, procedures associated with decision making in *games* have been developed. It is usual to express the decision outcomes of the game participants as numerical values.

In games, two or more decision makers, considered to be equally informed and intelligent, are placed in competition with each other. They are referred to as players, and their conflicts or competitions are referred to as games. The first step in a game situation is to examine the possible outcomes in the game. The assumption is made that the outcomes are clear and each player has a consistent preference for each outcome.

Alternative courses of action (decisions) are called strategies. The simplest example of games is the *two-person zero-sum game*, which is considered here. Zero sum implies that one person's gain is the other's loss. Games with more than two players are essentially similar but become very complicated quickly. Multiplayer games can be represented as two-person games by either of the following methods:

- One side of the game can be represented by the most aggressive member of the opposing players
- The entire array of competitors can be represented as a single opponent

12.4.1 Game Examples

There are two types of two-person zero-sum games. In one, the preferred position for each player is achieved by adopting a single strategy. This type of game is a *pure-strategy* game. The second type of game results in the players adopting a mixture of different strategies in order to achieve the preferred outcome and is therefore referred to as a *mixed-strategy* game.

Example: Pure Strategy

An example payoff matrix for a two-person zero-sum game is

		Player B Choices				Row Minimum
		B_1	B_2	B_3	B_4	
Player A choices	A_1	8	12	7	3	3
	A_2	9	14	10	16	9
	A_3	7	4	26	5	4
Column maximum		9	14	26	16	

By convention, a positive payoff means a gain for the row player (or the maximizing player) and loss to the column player (or the minimizing player). For the game given above, the optimal strategy for player A is identified by the maximin outcome while the optimal strategy for player B is identified by the minimax outcome. This is because each player will choose a strategy, which cannot be upset by the competitor.

- By choosing A2, player A is placed in a preferred position in that the payoff of 9 cannot be changed by B.
- Likewise by choosing B₁, player B is assured of an outcome that cannot be upset by the competitor.
- For example, if player A plays A2, any strategy other than B₁ will leave player B worse off. Similarly, if player B plays B₁, any strategy other than A2 will leave A worse off.
- The maximum for A is the same as the minimum for B. This is the point (A_2, B_1), which is known as a *saddle point*, and this is a *pure-strategy game*.

Example: Mixed Strategy

Consider a situation in which two companies are competing to secure a contract. Each company has three strategies (low bid, medium bid, high bid). The strategies and payoffs can be represented as a two-person zero-sum game.

		Company B Bids				
		B_1	B_2	B_3	Row Minimum	
Company A bids	A_1 (low)	2	5	7	2	Maximin
	A_2 (medium)	−1	2	4	−1	
	A_3 (high)	6	1	9	1	
Column maximum		6	5	9		
			Minimax			

What is the optimal strategy for each company?

It is clear that the minimax does not equal the maximin. Therefore, no saddle point exists for this game (it is not a pure strategy game).

Firstly, from inspection of the payoff matrix, notice that for player B, strategy B_2 always gives a better result than B_3. That is, B_3 is *dominated* by strategy B_2. Therefore, B_3 can be dropped from the payoff matrix. After B_3 is eliminated, notice that A_2 is dominated by A_1. Hence A_2 is eliminated. The payoff matrix then becomes

	B_1	B_2	Row Minimum
A_1	2	5	2
A_3	6	1	1
Column maximum	6	5	

There is no saddle point.

The mixed-strategy game can be dealt with in several ways, including graphical and linear programming methods.

12.5 Decision Trees

A visual method for comparing decisions is available through the use of decision trees. Decision trees are useful when the number of courses of action or decisions is not large and the number of possible states of nature is not large. They are useful in conjunction with EMV or expected utility value (EUV) for deciding the best sequence of decisions.

Decision trees are constructed according to the order followed in making a series of decisions. A tree consists of sets of alternatives, their outcomes, and corresponding probability assignments. A tree is read and developed from left to right as the decision process goes. Calculations, however, are done from right to left.

The tree is made up of decision nodes, chance nodes, and terminal or outcome nodes. Drawing the tree requires the identification of alternative decisions and the outcomes associated with these decisions.

Each decision tree starts with a decision node from which emanate alternatives. For each alternative, there may be several possible outcomes, shown branching from chance nodes. Each outcome has an associated probability of occurrence, typically based on informed opinion. The gathering of additional data (experimenting) may take the place of an outcome, and this leads to further alternatives, and so on. The outcomes at any chance node are mutually exclusive and collectively exhaustive, with their probabilities summing to 1. A useful reference on the topic is Ang and Tang (1984).

Symbols commonly used in decision trees follow Figure 12.1.

Calculations on decision trees indicate a sequence of decisions contingent on all possible outcomes. It is not a single sequence of decisions best for all

☐	A decision node. The point at which a decision is made.
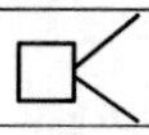 ☐<	Possible courses of action or decisions are represented by branches or action lines.
 ○	Chance node or event. Each action terminates in a chance node.
 ○<	The alternative outcomes for each chance node are represented by lines leaving the chance nodes. The values of the probability of each outcome occurring is noted on the branch.
△	Outcome, or terminal node, following a particular decision and chance node.

FIGURE 12.1
Decision tree symbols.

occasions. The trees provide a structuring and decomposition into important elements and provide understanding behind decisions. Uncertainty is still involved, but it allows the decision maker to be better equipped to deal with the uncertainty.

The order of magnitude, rather than the accuracy, of the numbers that come out of a decision tree is important. Sensitivity calculations should always be carried out. As well, the systematic structuring of the decision process, and the rigor this forces on the decision maker, is important.

12.5.1 Determining EMV

EMV is calculated as the probability of occurrence of an event times the monetary value of that event, summed over all associated outcomes:

$$EMV = \sum_{i=1}^{n} x_i P(x_i)$$

where
x_i, $i = 1, 2, \ldots$ is the monetary value of an outcome
$P(x_i)$ is the associated probability of occurrence

EMV represents the value that could be expected on average if the same decision was made in many similar situations. The calculations are done from right to left in the diagram.

Once the decision tree has been established, the EMVs of each node can be determined using the following sequence:

1. Start at the outcomes (ends of tree branches) at the right-hand side of the tree. Compute the EMV for each outcome.
2. Compute the EMV for chance nodes closest to the outcomes. If one chance node leads to another, use the EMV for the later occurring chance node as the expected value for the next branch.

3. At the first level of decision nodes,
   a. Compare the EMVs of the alternatives.
   b. Choose the best.
   c. Indicate which of the others will not be considered further—cross or mark these out.
   d. Assign the EMV of the best alternative to the decision node.
4. Proceed to the next level of chance nodes and repeat steps 3 and 4.
5. Continue the process until the root of the tree is reached.
6. Alternatives that are not crossed/marked out give the best decision based on EMV.

Example

A company needs to erect a temporary access bridge in an area of known seismic activity. It must decide what degree of earthquake resistance should be incorporated in the design to minimize total costs associated with seismic effects and damage. The following information is available:

- In 50 years of earthquake records, 40 earthquakes have occurred.
- To make the bridge earthquake resistant, the cost is as follows:

Earthquake Magnitude	Number in Past 50 Years	Cost of Resistant Design ($)
0–4	24	1000
4–7	12	3000
>7	4	6000

- The cost of damage if an earthquake occurs is thus:

Earthquake Magnitude Designed for	Earthquake Magnitude Occurring	Cost of Damage ($)
None	0–4	3000
	4–7	5000
	>7	8000
0–4	0–4	0
	4–7	3000
	>7	7000
4–7	>7	4000
>7	>7	0

There is only one level of decision—that is, what level of earthquake to design for. The tree is drawn from one decision node with four possible branches.

At the end of each branch, there is a chance node with two outcomes: an earthquake (probability 0.8); no earthquake (probability 0.2).

If an earthquake occurs, there are three possible ranges of magnitudes with the following probabilities: P(0–4) = 0.6, P(4–7) = 0.3, P(>7) = 0.1.

The decision tree is shown in Figure 12.2.

The EMV of each node can be calculated as follows:

$$EMV(F) = 0.6(-3000) + 0.3(-5000) + 0.1(-8000) = -\$4100$$

$$EMV(G) = 0.6(-1000) + 0.3(-4000) + 0.1(-8000) = -\$2600$$

$$EMV(H) = 0.6(-3000) + 0.3(-3000) + 0.1(-7000) = -\$3400$$

$$EMV(I) = 0.6(-6000) + 0.3(-6000) + 0.1(-6000) = -\$6000$$

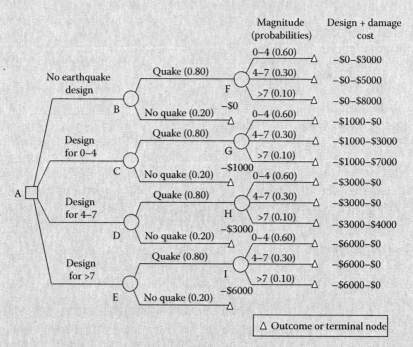

FIGURE 12.2
Decision tree for earthquake design example.

Now looking at the second lot of chance nodes,

$$EMV(B) = 0.8(-4100) + 0.2(0) = -\$3280$$

$$EMV(C) = 0.8(-2600) + 0.2(-1000) = -\$2280$$

$$EMV(D) = 0.8(-3400) + 0.2(-3000) = -\$3320$$

$$EMV(E) = 0.8(-6000) + 0.2(-6000) = -\$6000$$

This would indicate that the bridge should be designed to resist an earthquake of magnitude 0–4.

Example: The Effect of Different Contract Clauses

Owners can use decision tree ideas to assess the effect that different contract clauses may have on a project. Decision trees are, consequently, useful in the design of contracts, among other decision situations. For example, consider the case of deciding on what form the latent condition clause should take in a contract. Here it is assumed that the owner is contemplating using either a typical fair clause or not having a latent condition clause. In a more extreme situation, the owner might be contemplating the use of an onerous clause or non-onerous clause.

The owner has a choice of contract clauses. Irrespective of which contract clause is adopted, there is the possibility of a latent site condition arising. Should no latent site condition arise, then no further consideration is required. However, should there be a latent site condition, there are a number of possible outcomes. All of these outcomes have an associated cost from the owner's viewpoint (Figure 12.3).

A similar decision tree can be drawn from the contractor's perspective. The costs of the outcomes now become costs to the contractor rather than the owner. As well, the initial decision will be on whether to accept or alter the relevant contract clause—this decision being made at the time of tendering. Decision trees thus can assist the actions that contractors take, for example, in qualifying tenders and other tender decisions.

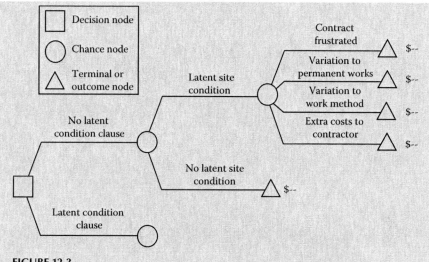

FIGURE 12.3
Decision tree (owner's perspective); decision on latent site condition clause.

Example: Contract Clauses and Disputes

Consider the example Figure 12.4 for disputes arising from different contract clauses.

At D,

$$EMV = 0.7(-\$0) + 0.2(-\$10{,}000) + 0.1(-\$20{,}000) = -\$4{,}000$$

The decision maker at C then has a choice of proceeding with a claim at an EMV of −$4000 or continuing without claiming at −$5000 (cost).

This process is repeated, successively comparing the EMV of the alternatives at each decision node. The nonpreferred paths are eliminated, again going from right to left. Figure 12.5 shows the nonpreferred paths struck out and the EMVs at the tree nodes.

At node A, it is found that the preferred path to take is to use contract clause version I and if a dispute arises to pursue the claim.

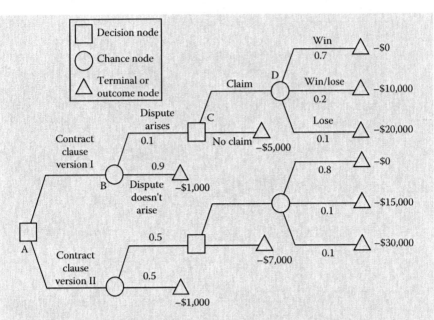

FIGURE 12.4
Decision tree; decisions and outcomes resulting from different contract clauses.

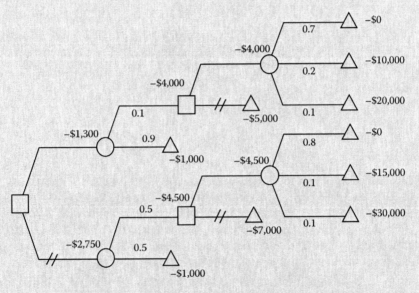

FIGURE 12.5
Decisions based on EMV.

Example: Which Dispute Resolution Method?

Each of the many possible contract dispute resolution methods has a number of advantages and disadvantages. As well, different cultures will be more relaxed with some methods of dispute resolution than with others. A party in one country may be reluctant to accept the law of another country; hence there may be a need to write an alternative dispute resolution method into international contracts with such parties.

When selecting a dispute resolution method, a decision tree approach may be helpful. Figure 12.6 shows a possible general structure, but specific projects and disputes would benefit from a more specific tree tailored to the particular situation.

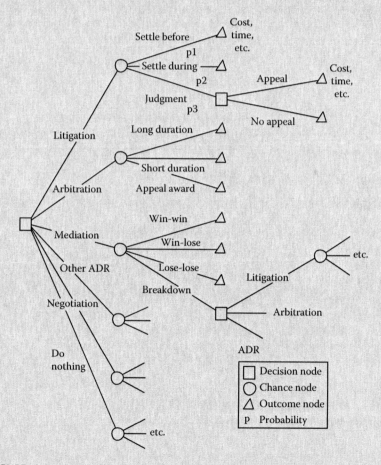

FIGURE 12.6

General dispute resolution decision tree. (After Carmichael, D.G., *Disputes and International Projects*, A.A. Balkema, Rotterdam, the Netherlands, 2002.)

The probabilities will typically be based on informed opinion. In Figure 12.6, the probabilities p1, p2, and p3 sum to 1; p1 is the probability that the matter will settle before reaching court, p2 is the probability that the matter will settle some time during the court hearing, and p3 is the probability that there will be a judgment (which could be further divided into a favorable judgment and an unfavorable judgment). Outcomes will be in terms of cost (both direct and indirect), but could also be in units other than money, for example, in terms of work hours spent. Using expected cost (or expected utility) would indicate the most appropriate decision (choice of dispute resolution method) to make.

Adopting such an approach gives informed decision making as to the most appropriate contract dispute resolution method to adopt. It removes much of the subjectivity commonly involved in making such a decision. Of course, the estimated probabilities could be manipulated by someone wishing the tree to indicate a particular resolution method.

12.6 Bayes Theorem and Additional Information

The previous examples have used estimates of probabilities arising in the decision process. One possible decision, applicable in many situations, is to hold off making a decision and obtain more information, for example, via experimentation. The new information allows improved estimates of the probabilities. Improved estimates of the probabilities can be obtained by applying Bayes theorem, which gives the conditional probability of two related events.

If there are two events, A and B, such that the occurrence of one event influences directly the occurrence of the other. The conditional probability of A occurring given that B has already occurred is given by

$$P(A|B) = \frac{P(B|A)P(A)}{P(B)}$$

Here, the conditional probability $P(x|y)$ reads as the probability of x given the occurrence of y, or conditioned on the occurrence of y.

If there are many events $X_1, \ldots, X_n$ that are possible outcomes of a decision with initial estimates of probabilities $P(X_1), \ldots, P(X_n)$, and an experiment that leads to an event E occurring is performed,

$$P(E) = \sum_{i=1}^{n} P(E|X_i)PX_i$$

and

$$P(X_i | E) = \frac{P(E|X_i)P(X_i)}{\sum_{i=1}^{n} P(E|X_i)P(X_i)}$$

12.7 Utility

An alternative to EMV is to use utility (expected utility value—EUV) as the basis of choice. EMV implies that every dollar gives the same amount of satisfaction irrespective of the total involved. Utility attempts to take into account personal differences in the value of money at different amounts.

Utility arises because different people generally place different values on the same thing. They exhibit different preferences. Their preferences are also usually nonlinear. EMV is linear and is not always suitable for representing the way people behave in decision situations.

Example

The following example demonstrates this.
There are two decision alternatives:

A_1 [You get $10,000], or

A_2 [You get $22,000 if, when a coin is flipped, tails is the result.

If heads is the result, you get nothing.]

Then,

$EMV(A_1) = \$10{,}000$

$EMV(A_2) = (1/2)(\$22{,}000) + (1/2)(\$0) = \$11{,}000$

By EMV, you would choose A_2.

Consider also the following alternatives:

B_1 [You lose $500], or

B_2 [You lose $10,000 with a probability of 1/100.

You have 99/100 chance of losing nothing.]

Then,

$$EMV(B_1) = -\$500$$

$$EMV(B_2) = (1/100)(-10,000) + (99/100)(0) = -\$100$$

By EMV, you would choose B_2.

Most people when faced with the earlier decision choices would choose A_1 and B_1 because, with A_2 and B_2, there is the possibility of losing large amounts. This is known as *risk aversion*. The term "risk aversion" is used when the decision maker is unwilling to accept EMV because s/he may lose considerable sums of money. The term *risk acceptance* is used when the decision maker takes an action that is not suggested by EMV because s/he stands to gain large monetary rewards (for example, in a lottery).

This example suggests that in some cases, an alternative to EMV should be used in decision making. One alternative involves the decision maker trying to optimize the EUV rather than the EMV. Using the notion of utility, the decision maker transforms the monetary values into nondimensional units, which reflect the decision maker's preference, which is usually nonlinear. This transformation is known as a *utility function*.

12.7.1 Methods of Utility Function Determination

12.7.1.1 Certainty Equivalent Method

To construct a utility function, identify all possible outcomes and rank them from most favorable (O^*) to least favorable (O_*). Give the most favorable a utility value of 1 and the least a utility value of 0. The decision maker then has to determine a situation in which two actions, a_1 and a_2, are possible.

a_1 has outcomes O^* and O_*, and a_2 has an outcome between O^* and O_* and a utility value between 1 and 0.

The decision maker needs to find a value of the probability p, which makes a_1 and a_2 equally acceptable. Once this probability (p) is found, the utility of an outcome O_i can be calculated using

$$\text{Utility of } O_i = U(O_i) = pU(O^*) + (1-p)U(O_*)$$

where $U(O^*) = 1$ and $U(O_*) = 0$ in most cases.

This process is repeated for other O_i, and the points so found describe the utility function.

Some experiments have demonstrated that measurements made using this method may be distorted, due to the probabilities used in assessment.

12.7.1.2 Example of Utility Function Determination

Two methods of utility function determination using certainty equivalents are demonstrated here:

The first gives dollar values and asks the decision maker to choose p.

The second gives probabilities and asks the decision maker to choose dollar values to achieve indifference.

12.7.1.2.1 Method 1: Given $, Choose p

Outline: A person is asked to make decisions involving consequences ranging from a $15,000 loss to a $30,000 gain. The utility of 0 is given to the loss of $15,000 and 100 to the gain of $30,000. The following lotteries are used.

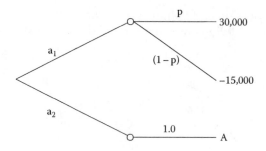

To construct the utility function, the value of A is varied throughout the range of values. The decision maker is then asked to decide on the probability that makes decisions a_1 *and* a_2 *equally acceptable.* The value of the chosen probability, p, is likely to be different for each person and reflects their individual preference.

Consider the following situation put to the decision maker.

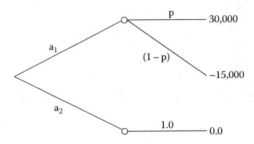

Say p = 0.8, and using

$$U(O_i) = pU(O^*) + (1 - p)U(O_*)$$
$$U(0) = (0.8)(100) + (0.2)(0) = 80$$

This gives a point (0, 80) on the utility function.
Consider now the following situation put to the decision maker.

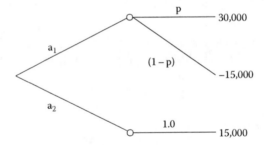

Say p = 0.9,

$$U(15,000) = (0.9)(100) + (0.1)(0) = 90$$

This gives a point (15,000, 90) on the utility function.
Consider now the following situation put to the decision maker.

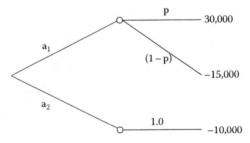

Say p = 0.3,

$$U(-10,000) = (0.3)(100) + (0.7)(0) = 30$$

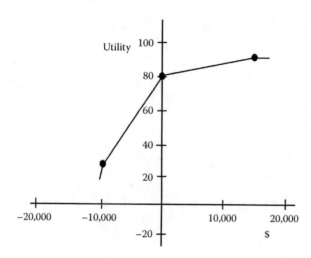

FIGURE 12.7
Utility function for example lottery.

This gives a point (–10,000, 30) on the utility function.

The three points calculated can then be plotted on a graph to give the utility function (Figure 12.7).

12.7.1.2.2 *Method 2: Given p, Choose $*

Example: A person has the opportunity of gaining $100 or a potential loss of $20. Establish a utility function over this range of dollar values.

Set a utility value of 1.0 to $100 and 0 to $0

Present the decision maker with the following decision situations.

First decision situation:

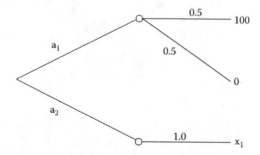

What value of x_1 would make a_1 and a_2 equally acceptable?
Assume $x_1 = \$30$ is chosen, then

$$U(30) = 0.5U(100) + 0.5U(0) = 0.5$$

This gives a point (30, 0.5) on the utility function.

Second decision situation:

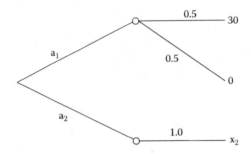

Assume $x_2 = \$15$ is chosen, then

$$U(15) = 0.5U(30) + 0.5U(0) = 0.25$$

This gives a point $(15, 0.25)$ on the utility function.
   Third decision situation:

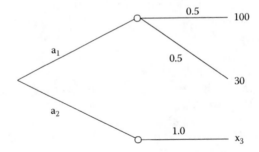

Assume $x_3 = \$55$ is chosen, then

$$U(55) = 0.5U(100) + 0.5U(30) = 0.75$$

This gives a point $(55, 0.75)$ on the utility function.
   Fourth decision situation:

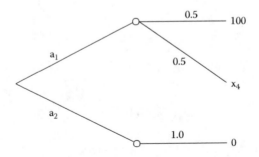

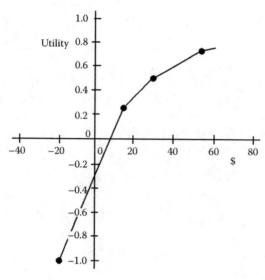

FIGURE 12.8
Utility function obtained in example.

Assume $x_4 = -\$20$ is chosen

$$U(0) = 0.5U(100) + 0.5U(-20)$$

$$U(-20) = 2(U(0) - 0.5U(100)) = -1.0$$

This gives a point (–20, –1.0) on the utility function.
And so on for other decision situations to give Figure 12.8.

12.7.1.3 Human Behavior

Utility function determination is a difficult task because it is subject to all
the vagaries of human behavior. Each person has a different utility function,
and the replies people give when questioned about utility can be influenced
by a number of factors. Care needs to be taken therefore in developing utility
functions.

There are a number of arguments both for and against the use of utility. One
argument against is associated with the measurement of utility. Because the
measurement is based on hypothetical lotteries, the elicitation of utilities may
take the decision maker away from the real world of the decision. However,
utility does not attempt to describe the way in which people make decisions.
Rather it indicates what a rational decision maker would do. In engineering
projects with a high level of uncertainty, utility has a valuable role to play for
decision makers who are familiar with the concept of probability.

Example

To demonstrate the different decisions that may arise from considering EUV compared with EMV, consider the decision situation of Figure 12.9. Here two decisions are possible: A and B. Associated with each decision are two possible outcomes with their respective probabilities shown in Figure 12.9.

The assumed utility function is Figure 12.8.

Expected monetary value calculation

$$EMV(A) = (0.2)(30) + (0.8)(15) = 18.0$$

$$EMV(B) = (0.6)(55) + (0.4)(-20) = 25.0$$

That is, B is the more preferable decision.

Expected utility value calculation

$$EUV(A) = (0.2)(0.5) + (0.8)(0.25) = 0.30$$

$$EUV(B) = (0.6)(0.75) + (0.4)(-1.0) = 0.05$$

That is, A is the more preferable decision.

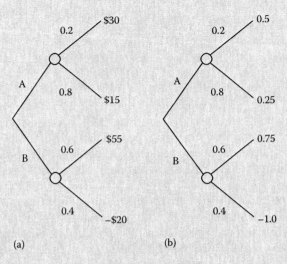

(a)   (b)

FIGURE 12.9
Example expected monetary and utility values. (a) Money, (b) utility.

12.7.2 Approximation for Utility Functions

It is possible to fit a function to utility values. This may improve computations with the utility function. Two popular forms suggested are as follows:

- The exponential function

$$U(X) = a - be^{-cX}$$

- The power function

$$U(X) = a + bX^c$$

Both can be considered reasonable approximations and are fitted typically using the two end points and a known intermediate point.

Example

Consider a range of values:

0 with utility 0
500 with utility 0.7
1000 with utility 1.0

Substitute the three known points

$$U(0) = 0 = a + b(0)^c \tag{12.1}$$

$$U(500) = 0.7 = a + b(500)^c \tag{12.2}$$

$$U(1000) = 1.0 = a + b(1000)^c \tag{12.3}$$

The first equation gives a = 0
  Divide Equation 12.2 by Equation 12.3

$$0.7 = \left(\frac{500}{1000}\right)^c$$

or

$$c = 0.514$$

Then

$$b = \left(\frac{1}{1000}\right)^c$$

or

$$b = 0.0287$$

This gives a function approximation of

$$U(X) = 0.0287X^{0.514}$$

Appendix A: Sensitivity

Sensitivity looks at the change in the system output or response resulting from changes in the following:

- The input: The input is varied a small amount away from some anticipated or usual value.
- The system makeup including system parameters, boundary conditions, initial conditions, and so on.

In both cases, the output is determined, and the magnitude of any change in the output is noted. As such, sensitivity belongs within the analysis configuration.

Commonly, it might be called a *what if* analysis. For example, *What if the input is changed* or *what if a system parameter is changed*? But strictly, sensitivity analysis only deals with *small changes* (± a few percent) to the input or system makeup (while "what if" changes can be anything at all). For such changes, the response is examined and compared with the original response. This comparison establishes whether the system is sensitive (to input changes or system makeup changes). Output changes of magnitude bigger than the perturbing changes or of an order of magnitude bigger than the existing output would commonly imply high sensitivity. Small magnitude changes in the output would imply low sensitivity. That is,

Sensitive: small change (in input or makeup) leads to large change in output.

Insensitive: small change (in input or makeup) leads to small change in output.

Typically, one input or parameter is varied at a time, and the other inputs and parameters are kept constant, in order to isolate the impact of that input or parameter.

Sensitivity analysis has importance if it is acknowledged that the real world is not deterministic and changes do occur in the environment (natural), material properties, operations, etc. As well, errors occur.

As with all results in analysis, they may be turned around to assist with investigation.

In terms of synthesis, it may be important to be able to design a system with low sensitivity.

Example: Risk Management

Sensitivity analysis is a major tool of risk management, because it is a deterministic way of acknowledging variability, and it allows people to avoid more difficult probabilistic analyses.

Exercise A.1

Consider the present value of $1 that you might receive in 1, 10, 50, and 100 years. The relevant compound interest formula (model) given in most economics/investment texts is

$$P = \frac{S_n}{(1+i)^n}$$

This gives the present worth (PW) equivalent (P) of an amount S_n occurring at year n, at interest/discount rate i. $1/(1 + i)^n$ is termed the present worth factor.

Do this calculation for interest/discount rates, i, of 1%, 5%, 10%, and 25% per annum, and years, n, of 1, 10, 50, and 100. Use the following table:

	1 Year	10 Years	50 Years	100 Years
1%				
5%				
10%				
25%				

What do you conclude about the sensitivity of present worth to changes in interest rate, i, and changes in years, n?

Exercise A.2

How good are our estimates of interest rates, costs, benefits, and asset lifetimes when predicting 1, 5, 20, and 50 years ahead?

Hence, how usable can a recommendation for investment in an asset be, based on assumed interest rates, costs, benefits, and lifetimes?

Example: Investment Feasibility

Investment decisions and background regarding the economic feasibility of investments (or projects) are generally based on indicators including the following:

- Present worth (PW) or net present value (NPV)
- Benefit: cost (B/C) ratio
- Internal rate of return (IRR)
- Payback period (PBP)

In performing analysis, assumptions are required, for example, on the following:

- Future interest/discount rates. This may reflect both national and international conditions.
- The lifetime of the asset/facility/end-product. This may depend on the multitude of forms of obsolescence.
- Future benefits resulting from the asset/facility/end-product. This may depend on future markets.
- Future costs of the asset/facility/end-product. This may depend on future usage, operation, and maintenance.

As well, there may be a need to make assumptions on the following:

- Inflation. This may reflect both national and international conditions.
- Depreciation of the asset/facility/end-product. This may depend on future usage, operation and maintenance, and taxation policies.
- Taxation implications. This is subject to government whims.

The assumptions of such analyses are not immune from GIGO (garbage in, garbage out) issues. Future interest rates, benefits, costs, and so on are extremely difficult to predict, and hence caution is suggested in imbuing the analysis results with too much accuracy. All results should be tested for the range of their validity. A sensitivity analysis assists with this.

A sensitivity analysis looks at varying these assumptions (input) by small amounts and examining the change in the economic indicators (output) of PW, B/C, IRR, and PBP. Typically, one assumption is varied at a time and the other assumptions are kept constant.

When comparing two investments, one investment may be more preferable under one range of assumptions, while the other investment may be more preferable under another range of assumptions. It is possible in these circumstances to find a point where the two investments become economically equal. Such a point is referred to as the *break-even point*.

Considering some example values, Figure A.1 shows how present worth varies with interest rate for the example values.

The interest rate that gives a PW of zero is about 15%. The investment is acceptable at any interest rate below 15%, but not above 15%.

Considering some different example values, Figure A.2 shows some results of a sensitivity study of the effect of a change in assumptions on IRR for the example values.

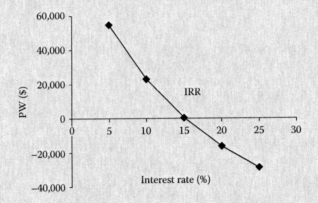

FIGURE A.1
Example change in PW with interest rate.

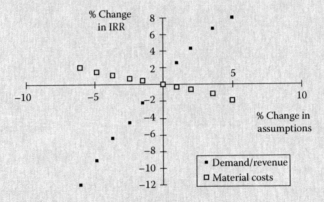

FIGURE A.2
Example change in IRR with change in values of some assumptions.

Exercise A.3

What is the current interest rate at which you can borrow money from a bank or financial institution? At what rate will you be able to borrow next year, and the following 9 years after that? And hence how would you use a sensitivity analysis for the analysis of the impact of changes in interest rates?

When borrowing money to buy a house, the loan can either be at a fixed interest rate for a given number of years into the future, or you can choose to have a variable interest rate that fluctuates with the market. How do you decide what is the better deal for you? How does the bank, from which you are borrowing, determine its fixed interest rates?

Exercise A.4

Considering the topic of sensitivity analysis, after having established all the influences affecting the setting of interest rates, what rate (today's rate plus and minus) would you use in a sensitivity analysis to examine the effect of interest rates on financial calculations into the future (for example, compound interest, present worth, benefit: cost ratio, payback period).

Note that the change in interest rate is not the risk. It is the risk source. The risk is related to the extra (or less) cost to you resulting from the change in interest rate. You might even go back further to establish an even more fundamental source of risk, rather than interest rate change—for example, politicians' mishandling affairs, global financial crisis (GFC), or regional unrest.

Most people say that future interest rates are unpredictable to any degree of exactness. This means that present worth values of future benefits and future costs are less exact the further into the future that these benefits and costs occur. (Is higher risk associated with higher uncertainty?) But discounting formulae (essentially compound interest) (particularly for large interest rates) give less weight further into the future. Do these two things then cancel themselves out? That is, although uncertainty increases, present worth magnitude decreases, and hence does this give a constant risk level over time?

Example: Earthmoving

A sensitivity analysis for an earthmoving, quarrying, or surface mining operation involves observing how the output changes with changes in its input or parameters such as

- Truck number
- Truck/scraper capacity
- Loader/dozer capacity
- Haul route length

The operation is altered by plus and minus a small value in each of these inputs or parameters in turn, while keeping all other variables unchanged and as perhaps observed in the field. That is, one input or parameter is varied at a time. Typically, the effect on production, cost per production, or emissions per production of a change in these inputs or parameters will be of interest.

Lower travel times can be brought about by

- Improving the haul roads' surface condition (rolling resistance, RR) through the use of a grader, water cart, and compactor
- Removing bends in the haul roads and superelevating bends
- Reducing grades (grade resistance, GR), or travelling loaded downhill and unloaded uphill

Varying the truck/scraper size will change the load time and travel time.

Varying the loader bucket size will change the load time. Varying the dozer (in the dozer-scraper case) capacity will have a similar effect.

Varying the bucket numbers will change the load time. With lighter/heavier loads, the travel times and possibly dump times will also be affected.

Appendix B: Surveys

B.1 Research

Research relies on data, and sometimes these data cannot be obtained through experiment or observation.

Such nonexperimental or nonobservational data can be categorized as primary or secondary. Primary data refer to data collected specifically for the investigation at hand. Secondary data are data that may have been compiled for some other purpose, and by others, yet are still usable. These secondary data may be available free of charge or may be available for purchase.

Secondary data sources include libraries; government publications about, for example, company details, census data, incomes, expenditures, forecasts, ...; trade, professional and business associations; companies; and universities.

Primary data may be collected in a number of ways by surveys including personal and mail interviews, email, Internet based, telephone; observation; and experiment.

B.2 General Comment on Surveys

To many people, surveys are research. They are the predominant tool used in marketing and management research. The reasons for this usage are

- Cheapness
- Ease of carrying out
- Ability to measure thoughts and attitudes
- The restricted backgrounds of the researchers preventing alternative tools being used

In a political context, a survey might be referred to as an *opinion poll*.

Surveys are a fast and inexpensive means of obtaining research information. However, there are many pitfalls to conducting surveys. It is not an easy task designing a good questionnaire, as well as encouraging truthful and accurate replies.

Surveys can be viewed in a negative light. There are many well-known pitfalls with surveys (selection bias, low reply rate leading to skewed results, encoding and decoding of questions, and so forth), and it is a real challenge for those conducting surveys to eliminate these factors.

While survey results should not be interpreted in isolation or used as the sole source of information, they can be of use in combination with other approaches and can be usefully applied to many situations.

Although most surveys are not done well, it does not invalidate the argument that a good survey can provide valuable information.

Example

A long (considered boring by a respondent) survey started off with general questions and then became more specific as it got further into the survey. The survey asked the same questions in slightly different ways and to slightly different depths. By comparing replies to this survey, patterns and trends in the answers could be seen.

A lot of so-called research in the field of management has not progressed beyond surveys. A major reason for this is the lack of rigorous quantitative backgrounds of the "researchers." Nevertheless, these people see it as valid research even though the surveys and results can be easily dismantled.

From an academic research viewpoint, it is extremely difficult to get rigor from a survey. In marketing, this may not matter: How good are the predictions of election outcomes? How good are the audience figures published for television programs? How good are the origin-destination surveys and similar in predicting vehicle movements on proposed roads? How well can you predict consumer or industrial demand? Etc.?

Interpreting surveys is just as tricky as designing a survey and sampling. It is always amusing when a radio station publishes results of listener telephone polls. For example, the radio station may say that in a telephone poll, 50% of the listeners voted yes. Presumably two people phoned? This is the territory of the book, *How to Lie with Statistics* (Huff, 1954).

Exercise B.1

Based on your personal experience, how useful do you believe are the results of surveys?

Can you ever design a survey that cannot be criticized for some reason?

Example: Marketing

There are two types of surveys that pretend to be market research but are not. They give market research a bad name and can turn people off contributing to genuine market research surveys. The television program Horizon referred to these as "suggers" (selling under the guise of research) and "fruggers" (fund raising under the guise of research).

Exercise B.2

How might you deal with the general fishing expedition used by researchers, where the researcher has not stated any hypothesis, but just wants to find out something (not defined)? Or do such ill-defined approaches not work within the scientific research approach?

Exercise B.3

How different, in terms of pitfalls, is soliciting information/feedback, say through discussions, to written surveys? Can oral discussions be "steered" to solicit a biased reply?

Exercise B.4

How limited are surveys in their application? For example, how useful are surveys for finding out about potential human resources issues in the workplace?

B.3 Questions

Questions raised in surveys/questionnaires are chosen to uncover particular pieces of information.

Closed-end questions of the form

Indicate your preference:

    []   A
    []   B
    []   C

tend to be preferred over open-end questions such as

> What is your preference?

Open-end questions may be useful for uncovering information previously not considered. That is, they can be used in exploratory and pilot studies. On the other hand, boxes are easier to check than writing answers, the results of checked boxes are easier to reduce and compare, and open-end questions may be interpreted by the respondent in a way not useful to the survey and may depend more on the interest of the respondent.

Open-end questions can work if the respondents are committed. They might also be used at the end of the survey as a catch-all question, allowing respondents to "have their say."

> **Exercise B.5**
>
> It is believed that the best respondents are "average"-type people who work and have a high school education. Very well educated people are believed to be poor respondents, because they try to double guess the survey, and are critical of the format, style, etc., of individual surveys.
>
> Does your experience support such a belief? If so, how does this affect the results of a survey?

B.4 Some Suggestions

Some suggestions for good practice in developing surveys are thus (Lehmann, 1979):

- Phraseology—be direct; avoid slang and fancy/polysyllabic wording
- Include provision for don't knows and refusals where appropriate
- Ask simple, not complex questions
- Use nontechnical language
- Group questions of similar type together
- Use mutually exclusive and exhaustive categories for multi-choice questions
- When appropriate, use nonsubjective rather than subjective scales for key constructs

- Avoid unbalanced scales
- Rate items attribute by attribute
- Be careful of the order in which a set of alternatives is presented
- Don't worry too much about the physical format for scaled questions
- Try to get relative weightings

Exercise B.6

In political elections, there is the so-called donkey vote where people number the ballot card in the order in which the candidates appear. Also the first box gets a disproportionate number of first choices.
  How can this effect be overcome in a survey?

Consideration also needs to be given to

- The order in which the questions are placed
- The placement of prying/offensive or experimental questions
- The variety of the format
- Branching of questions (questions that follow from earlier questions)

Some Typical Errors in Questionnaire Design (Stanton et al., 1985)

- The respondent feels the information requested is none of your business: What is your family's income? How old are you? What percentage of your home mortgage remains to be paid?
- Questions lack a standard of reference: Do you like a large kitchen? (What is meant by "large"?) Do you attend church regularly?
- The respondent does not know the answer: What is your spouse's favorite brand of ice cream?
- The respondent cannot remember and, therefore, guesses: How many calls did you (as a sales rep) make on office supply houses during the past year?
- Questions are asked in improper sequence. Save the tough, embarrassing ones for late in the interview. By then, some rapport has usually been established with the respondent. A "none-of-your-business" question asked too early may destroy the entire interview.

Example

A research survey, conducted by an external consultant for a company, quizzed the company's staff. One respondent's answers were based on

- Confusion
- Giving a definite yes/no answer when the respondent really meant "I don't know"
- A wanting for it to end
- Fear of job loss if the respondent gave a self-perceived wrong answer

The usefulness of the survey results would appear to be questionable.

Exercise B.7

For usefulness, do survey questions need to be written and structured in such a way that they tap into the subconscious mind of the respondent, in order to encourage reasoned, honest answers?

For survey usefulness, what guidelines would you suggest for matters such as survey size, groups of people, and types of questions?

What do you think of the practice of offering some token gift to encourage high reply rates?

What usefulness issues arise with ongoing surveys that are carried out on the same target population in order to understand trends or patterns?

The usefulness of survey results could be expected to be affected, among other things, by

- The respondents' knowledge on the topic
- The likely reply rate: who responds and the level of non-reply
- Potential for respondents to misunderstand the questions
- Faults and bias in the design of the survey: distortion in the results of the survey
- Sampling errors
- Errors in processing or statistical data reduction
- Faulty interpretation of results (also misinterpretation of body language, if applicable)

B.5 Survey Type

Surveys may be of a number of types:

- Personal (in home, office) interviews
- Telephone interviews
- Mail surveys
- Email, Internet based
- Drop-off, callback
- Panels
- Group interviews
- Location interviews

Drop-off callback contains parts of several survey types.

Panels involve a group of people agreeable to participating in a survey. They may record their information in diaries, they may be externally monitored, or agree to answer questionnaires. An example, familiar to many, of the second type is the meter that records television-watching habits. Each survey type has advantages and disadvantages (Lehmann, 1979).

B.5.1 Personal Interviews

Advantages

- Relatively complex presentations can be shown.
- Depending on what answer a respondent gives to a particular question, the interviewer can then branch to the next appropriate question.
- Respondents can be asked to give replies other than multiple-choice type.
- The presence of the interviewer can help convince the respondent to answer questions s/he might otherwise leave blank.

Disadvantages

- The presence of the interviewer may influence the replies.
- Since the interviewer often is required to interpret the reply and assign it to a predesignated category, there is serious potential for errors in interpretation.
- There is a strong possibility of interviewer cheating.

B.5.2 Telephone Interviews

Advantages

- Like personal interviews, phone interviews may follow fairly elaborate branching patterns.
- The phone interviewer may help "prod" the respondent to answer questions.
- Some individuals are more likely to answer a phone than to let a stranger enter their house/office/site.
- It can be completed quickly.
- It can be done from a single location.
- Interviewer bias is reduced somewhat.

Disadvantages

- The questions must be asked without any visual props.
- People may well respond differently.
- The replies are limited.
- As in the case of personal interviews, there is the opportunity for the interviewer to exert influence.

B.5.3 Mail/Email Surveys

Advantages

- The respondent is allowed to work at his or her pace.
- No interviewer is present.
- With adequate instructions, fairly complicated scales (for example, 10 points) can be used.
- By receiving the replies directly, the possibility of interviewer cheating is essentially eliminated.

Disadvantages

- There is no one present to prod the respondent to complete the questions.
- No one is available to help interpret instructions or questions.
- There is a nontrivial chance the survey will be treated as another piece of junk mail.
- Mail lists may be out of date, and people on the list may have moved.

- Most mail surveys are left open 3–4 weeks.
- Respondents tend to be slightly more upscale.

B.5.4 Panels

Advantages

- The reply rate among panel members is extremely high.
- For established panels, a great deal of other information such as demographics is already available.
- Using a panel is "easy."

Disadvantages

- At least in one aspect, panel members are clearly not typical individuals.
- Panels have a tendency to age.
- Most mail panels, including those designed to represent all segments of the population, tend to underrepresent both minority groups and low education levels.
- Being in a panel tends to "condition" respondents.

Exercise B.8

Apart from the earlier advantages and disadvantages of the various types of surveys, the types can also be compared on the basis of cost and reply rate. Comment on the suggested costs and reply rates in the following table:

	Cost	Reply Rate
Personal interview	Most expensive	50%–80%
Telephone interview	Relatively inexpensive	40%–75%
Mail/email survey	Relatively inexpensive	0%–60%
Panel	Medium expensive	70%–100%

Exercise B.9

Rank the different types of surveys according to the objectives of

- Speed
- Suitability for complex questions

Appendix C: Sampling

Sampling in surveys becomes necessary when it is impractical or too costly to obtain information on the complete population or universe. The key issue is choosing a sample, which is representative of the larger population, with respect to the matter being studied. Questions of size and type of sample then arise. In terms of sampling and using surveys, another key issue is non-reply or partial reply and dealing with the adjusted sample size.

Generally, the larger the sample size, the better the result. The uncertainty in the result decreases with sample size, but the cost increases.

The sample size is chosen to give a required degree of confidence in the statistics being calculated, whether these are sample mean (average), sample variance (standard deviation squared), confidence intervals, or other.

Sampling takes a number of forms, with the form adopted depending on the given situation.

C.1 Random Sampling

Where each item, person, … of the population (the whole group of people, practices, items, … being studied) has an equal probability or chance of being selected, the sample is termed random. Selection of the items, people, … in the sample is based on a set of random numbers available in tables or generated using a pseudo random number generator (numerical algorithm), much like throwing a dice or spinning a roulette wheel.

A variation on this is called *area sampling* where any area (involving a collection of people) has an equal probability of being chosen, but within the areas chosen, all people are surveyed.

C.2 Stratified Sampling

Where representation in a sample is wanted from different segments or strata of a population, the population is firstly stratified and then sampled. The alternative is random sampling of the population, and this possibly could lead to no representation from some segments.

Sample sizes of particular segments may be chosen in proportion to the segment size (proportionate sampling), chosen to give each item, person, … an equal chance of being selected (disproportionate sampling), or deliberately chosen in a disproportionate manner.

Quota samples are chosen to be proportional to some characteristic. They are accordingly not random but rather stratified or layered. Each item, person, ... does not have an equal probability of being chosen. Within a layer or stratum, however, each item, person, ... has an equal probability of being chosen.

> **Exercise C.1**
>
> Random sampling allows you to establish the accuracy of your results. How do you establish the accuracy of quota sampling? How big an influence does judgment play in quota sampling?

C.3 Other Sampling

Census—a universal survey where everything is surveyed.

Cluster sample—for convenience and cost reduction.

Sequential and replicated samples—where the first sample does or does not demonstrate something, a follow-up sample may be taken.

Multistage sampling—at each stage, a sample representing a different characteristic is drawn.

C.4 Non-Reply

Bias in a particular sample may be brought about through the original choice of the sample—a section of items, people, ... may not be included in the sample (noncoverage)—or through non-reply.

Non-reply may be brought about through refusals, inability to respond, and inability to find the respondents.

Whether a reply is obtained may depend on the survey characteristics and the respondent's view of the survey. Factors include

- Interest in the subject matter
- Length of the survey
- The survey opening/introduction
- Incentives to complete the survey

- Format of the survey
- Advance notice
- Callbacks, follow-ups

Exercise C.2

To allow for non-reply, it seems reasonable to adjust the results. This firstly requires estimating who the nonrespondents are. What would be a reasonable way to adjust the results?

Exercise C.3

Assume that a customer's probabilities of buying three brands (A, B, and C) are 0.6, 0.3, and 0.1, respectively.

a. Using a table of random numbers, simulate 10 purchases.
b. Simulate 10 more purchases.
c. How representative are (a) and (b) of the customer's true purchase behavior?

Exercise C.4

Assume that you had been retained to take a countrywide sample of 1000 to gauge opinions about a new product. Set up a scheme to do personal interviewing.

a. Use geographic regions as a starting point and draw two regions at random.
b. Use geographic regions as a starting point and draw two regions randomly with the probability of inclusion proportional to their population.
c. Which of (a) or (b) seems better?
d. How would you go about sampling within the geographic regions?

Exercise C.5

An email survey that you conducted only returns a reply rate of 40%. How does this introduce bias into the result?
  What steps could be taken to reduce this bias?

Exercise C.6

Vehicle registration lists, telephone directory lists, and similar are common sources of contacts for surveys.

How would you select a random sample from such lists?

Are such lists representative of the population at large? If not, what sectors are missing?

How do you correct for any bias because some sectors are missing?

Exercise C.7

You wish to study the usage of professional indemnity insurance by consulting engineers through interviewing a sample of 500 engineering practices drawn from those listed in the telephone directory under "Engineers-consulting." It is proposed to use a stratified sample "to save money," by sampling only those companies engaged in structural engineering. Comment on the rationale of such an approach.

Exercise C.8

Some studies in the past have shown that a small percentage of field interview results are fictitious—made up by the interviewer.

How do you correct the data reduction for this?

Studies have also shown that a much larger proportion of interview results contain errors. How do you correct the data reduction for this?

Exercise C.9

In looking to see what the market is for a new consumer product, what alternatives are available to a national survey? Discuss the suitability of the alternatives.

Appendix D: Measurement and Scales

In research, something is measured. This thing (sometimes termed a construct) may be an attitude, quantity, or other. People are more at ease working with quantitative scales for these constructs, although this might involve simplification. Being quantitative permits reduction and graphical display and also permits calculations to be performed.

Scales may be classified as

- Nominal
- Ordinal
- Interval
- Ratio

(See Figure D.1.)

The choice of scale may be dictated by the construct being measured. The scale, in turn, influences the form any subsequent data reduction takes.

The first two are nonmetric; the last two are metric. They are listed in terms of the number and types of reductions that can be performed on measurements done according to the scales; nominal scales permit only a few calculations while ratio scales permit all statistical calculations.

D.1 Nominal Scale

A nominal, cardinal, or categorical scale has no meaning in itself and does not relate to other scales. There is no relationship between the amount of the construct and the numerical measure (Figure D.1a). It denotes quantity but not order in a group.

For example, different colors may be allocated different numbers.

The only number manipulation that can be performed relates to working out the number of occurrences and frequencies. Statistical measures such as mean and standard deviation have no meaning.

Exercise D.1

Give an example of a nominal scale.

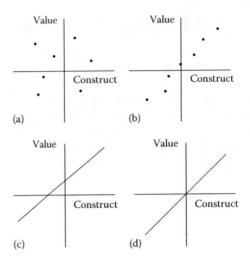

FIGURE D.1
Example scales. (a) Nominal, (b) ordinal, (c) interval, and (d) ratio. (After Lehmann, D.R., *Market Research and Analysis*, Irwin, Homewood, IL, 1979.)

[ ]	Answer A (=1)
[ ]	Answer B (=2)
.	
.	
.	
[ ]	Answer E (=5)

FIGURE D.2
Multiple-choice questions.

D.1.1 Multiple-Choice Questions

Multiple-choice questions (Figure D.2) generally imply a nominal scale when numbers 1, 2, … are attached to the different replies. The numbers bear no relationship to the quantity of the construct; they only indicate different reply categories.

The numbers 1, 2, … may be included in the survey, and the respondents asked to answer in terms of the numbers.

A special form of multiple-choice questions is the binary (yes–no) case where respondents answer yes (=1) or no (=2).

D.2 Ordinal Scale

In an ordinal scale, there is a relationship between the amount of the construct and the numerical measure, but not the absolute value of the measure (Figure D.1b). It denotes order, quality, or degree in a group such as first, second, …

Rankings are examples of ordinal scales—the higher the rank, the better. An item ranked first is better than one ranked second, but there is no indication as to how much better.

An ordinal scale permits medians and percentiles to be calculated.

Exercise D.2

Give an example of an ordinal scale.

D.2.1 Forced Ranking

Surveys where respondents are asked to rank alternatives (Figure D.3) lead to ordinal scales.

D.2.2 Paired Comparison

Paired comparisons ask the respondent to show preferences (Figure D.4), considering alternatives two at a time rather than all alternatives simultaneously.

Rank the following (1 indicating most preferred, 2 indicating second preference etc.):

    [ ]   Alternative A
    [ ]   Alternative B
    .
    .
    .
    [ ]   Alternative E

FIGURE D.3
Forced ranking.

Indicate the preferred alternative in each case:

Alternative A, Alternative B
Alternative A, Alternative C
Alternative A, Alternative D
Alternative A, Alternative E
Alternative B, Alternative C
Alternative B, Alternative D
Alternative B, Alternative E
Alternative C, Alternative D
Alternative C, Alternative E
Alternative D, Alternative E

FIGURE D.4
Paired comparison.

Question ....
Answer:

[]	[]	[]	[]	[]
Low	Below average	Average/ neutral	Above average	High
(=1)	(=2)	(=3)	(=4)	(=5)

Alternative

| (=−2) | (=−1) | (=0) | (=+1) | (=+2) |

FIGURE D.5
Semantic scale.

The number of times each alternative is preferred over another alternative gives the ordinal measures.

For 5 alternatives, there are 10 pairwise comparisons necessary; for 15 alternatives, there are 105 pairwise comparisons. This makes paired comparisons only viable for small numbers of alternatives.

D.2.3 Semantic Scale

Replies are asked for according to semantic categories (Figure D.5).

The intervals between categories cannot be related.

Constructs such as attitude might be measured by summing the answer to several questions, each testing one aspect that contributes to attitude.

D.3 Interval Scale

Where the differences between numerical measures, rather than the absolute values of the measures, are important, there is an interval scale (Figure D.1c).

An example is a Fahrenheit thermometer where the upper and lower limits of 32 and 212 have been set arbitrarily. Increases in temperature are indicated by higher numbers, but, for example, 100° is not twice as hot as 50°.

An interval scale permits means, standard deviations, parametric statistical tests, correlations, regression studies, discriminant "analysis," and factor "analysis," to be performed on the data. Only a few statistical tools cannot be used.

Exercise D.3

Give an example of an interval scale.

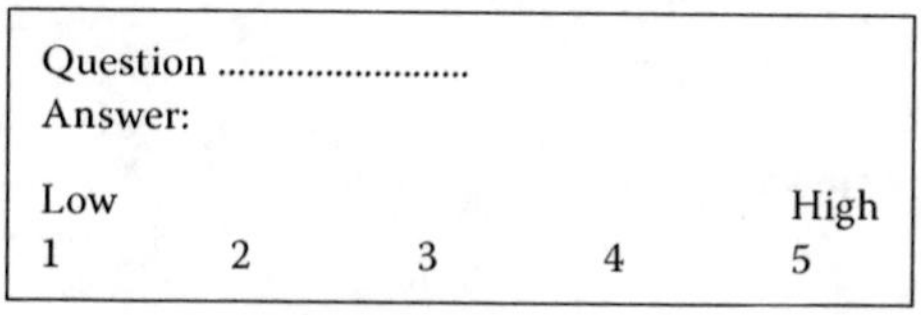

FIGURE D.6
Bipolar adjective, discrete scale.

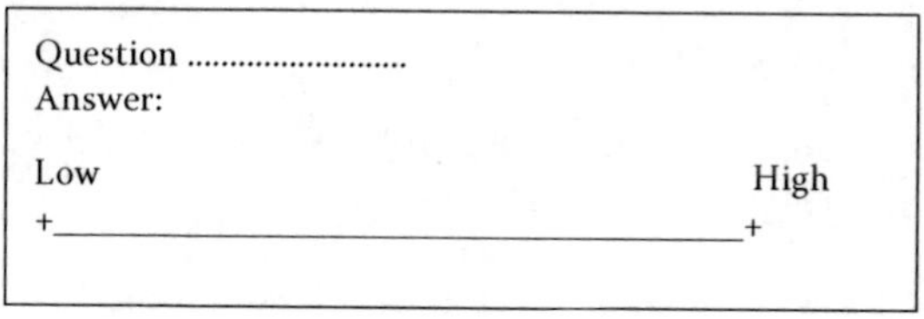

FIGURE D.7
Bipolar adjective, continuous scale.

D.3.1 Bipolar Adjective

For bipolar adjective, the same range is used as in a semantic scale, but the range is graduated equally between the lower and upper extremes (Figure D.6). Only the lower and upper categories are labeled.

Instead of discrete points between the lower and upper categories, a continuous scale might be used (Figure D.7).

Typical surveys have between four and eight scale points depending on the commitment and knowledge of the respondents. Telephone and like surveys use the lesser number of scale points.

Bias can be introduced if the scales are not balanced equally to the low and high ends.

D.3.2 Other Approaches

Other approaches use comparisons, typically in pairs.

D.4 Ratio Scale

A ratio scale implies meaningfulness to the absolute measurement values and the intervals between measurement values. As well, a value 0 implies the absence of the construct (Figure D.1d).

An example is money.

All statistical reductions may be carried out on measurements to a ratio scale.

Exercise D.4

Give an example of a ratio scale.

D.4.1 Direct Quantification

Surveys may ask directly for the actual amount of something.

D.4.2 Constant Sum Scale

A total number of points (for example, 10) are divided between a number of alternatives by the respondent. For example,

Alternative A	3
Alternative B	2
Alternative C	5
	10

Recalculation is necessary if the respondent gives values, which do not sum to 10.

A modification of the constant sum approach is the Delphi procedure.

Other approaches compare alternatives, for example, by leaving the constant sum value open to the respondent.

D.5 Comments

In using a particular scale, several matters raise their heads:

- Validity
- Consistency or repeatability
- Bias
- Efficiency
- Errors

Validity refers to the ability of a measure to represent a construct. Bias occurs when a result is obtained different to the true value. Efficiency is in terms of economy of effort to obtain a result.

How much faith can be placed on the results of a survey will depend on sample design, sample size, refusal rate, nature of the questions, and so on.

As the number of surveys proliferates by mail, telephone, email, and so on, refusal rates could be expected to rise, casting doubt on the usefulness of any survey results.

Table D.1 indicates some possible sources of errors.

TABLE D.1

Sources of Error

General Source	Type
Researcher/user	Myopia (wrong question)
	Inappropriate data reduction
	Misinterpretation
	Mistaken
	Researcher expectation
	Communication
Sample	Frame (wrong target population)
	Process (biased method)
	Reply (biased respondents)
Measurement process	Conditioning
	Process bias
	Recording
	Interpretation (mistaken)
	Carelessness
	Fudging
Instrument	Individual scale item
	Rounding
	Truncating
	Ambiguity
	Test instrument
	Evoked set
	Positional (order)
Respondent	Reply style
	Consistency/inconsistency
	Boasting/humility
	Agreement (yea saying)
	Acquiescence
	Lying
	Extremism/caution
	Socially desirable
	Reply
	Mistakes
	Uncertainty
	Inarticulation

Source: Lehmann, D.R., *Market Research and Analysis,* Irwin, Homewood, IL, 1979.

Exercise D.5

The forms of measurement in surveys are typically applied to individuals. What is the effect of aggregating individuals' replies to give an indicator of group behavior?

What is the influence of averaging?

Can groups be considered to be homogeneous?

How do you compare subjective matters such as attitude or satisfaction?

Exercise D.6

Respondents to a survey can give different replies depending on whether the probing is direct or indirect. In direct probing, respondents may hide their feelings or lack the ability to express themselves clearly.

If direct probing is used, how could more representative replies be obtained? For example, could you ask about a third person rather than the respondent?

Alternatively, if indirect probing is employed, how useful are approaches such as scenario interpretation (thematic apperception), word association, and ink blot tests as used in creativity style studies?

Exercise D.7

Hard data are those taken from published records and similar sources. Would you expect the usefulness and suitability of hard data and survey data to be different? Compare.

Exercise D.8

No matter how well measurement programs are devised and carried out for surveys, there will always be some random error. How can random errors be allowed for in the data reduction?

What happens if an error due to one source is cancelled out by an error due to another source? Are the results of any use?

As a person progresses through a survey, their replies may change due to conditioning or fatigue. How can this be taken into account in the data reduction?

Bibliography

Adams, S. (1997), *The Dilbert Future*, Harper Business, New York.

Ang, A. H.-S. and Tang, W. H. (1975), *Probability Concepts in Engineering Planning and Design, Vol. I, Basic Principles*, John Wiley & Sons, New York.

Ang, A. H.-S. and Tang, W. H. (1984), *Probability Concepts in Engineering Planning and Design, Vol. II, Decision, Risk and Reliability*, John Wiley & Sons, New York.

Antill, J. M. (1970), *Civil Engineering Management*, Angus and Robertson, Sydney, New South Wales, Australia.

Antill, J. M. and Farmer, B. E. (1991), *Antill's Engineering Management*, 3rd edn., McGraw-Hill, Sydney, New South Wales, Australia.

Ashby, W. R. (1956), *An Introduction to Cybernetics*, University Paperbacks, London, U.K.

Au, T. and Stelson, T. E. (1969), *Introduction to Systems Engineering—Deterministic Models*, Addison-Wesley, Reading, MA.

BBC-1 (1966), *Three Men on Class*, The Frost Report, April 7, 1966.

Beer, S. (1967), *Cybernetics and Management*, The English Universities Press, London, U.K.

Benjamin, J. R. and Cornell, C. A. (1970), *Probability, Statistics and Decision for Civil Engineers*, McGraw-Hill, New York.

von Betalanffy, L. (1971), *General System Theory*, Allen Laine The Penguin Press, London, U.K.

Bowen, J. (1987), *The Macquarie Easy Guide to Australian Law*, The Macquarie Library, Sydney, New South Wales, Australia.

Bryson, A. E. and Ho, Y.-C. (1975), *Applied Optimal Control*, John Wiley & Sons, New York.

Carmichael, D. G. (1979a), The state estimation problem in experimental structural mechanics, *Third International Conference on Applications of Statistics and Probability in Civil and Structural Engineering*, The University of New South Wales, Kensington, New South Wales, Australia, pp. 802–815.

Carmichael, D. G. (1979b), Optimal filtering of concrete creep data, *Seventh Canadian Congress of Applied Mechanics*, University of Sherbrooke, Sherbrooke, Quebec, Canada, Vol. 1, pp. 121–122.

Carmichael, D. G. (1980), Identification of cyclic material constitutive relationships, *Proceedings of the Seventh Australasian Conference on the Mechanics of Structures and Materials*, The University of Western Australia, Crawley, Western Australia, Australia, pp. 130–132.

Carmichael, D. G. (1981), *Structural Modelling and Optimisation*, Ellis Horwood Ltd (John Wiley & Sons Ltd), Chichester, U.K.

Carmichael, D. G. (1982), Adaptive filtering in structural dynamics, *Engineering Optimization*, 5, 235–247.

Carmichael, D. G. (1987), *Engineering Queues in Construction and Mining*, Ellis Horwood Ltd (John Wiley & Sons Ltd), Chichester, U.K.

Carmichael, D. G. (1989a), Production tables for earthmoving, quarrying and open-cut mining operations, pp. 275–284, in *Applied Construction Management*, Unisearch Ltd Publishers, Sydney, New South Wales, Australia.

Carmichael, D. G. (1989b), *Construction Engineering Networks*, Ellis Horwood Ltd (John Wiley & Sons Ltd), Chichester, U.K.

Carmichael, D. G. (2000), *Contracts and International Project Management*, A.A. Balkema, Rotterdam, the Netherlands.

Carmichael, D. G. (2002), *Disputes and International Projects*, A.A. Balkema, Rotterdam, the Netherlands.

Carmichael, D. G. (2004), *Project Management Framework*, A.A. Balkema, Rotterdam, the Netherlands.

Carmichael, D. G. (2006), *Project Planning, and Control*, Taylor & Francis, London, U.K.

Churchman, C. W. (1968), *The Systems Approach*, Delta Publishing Co, New York.

Churchman, C. W., Ackoff, R. K., and Arnoff, E. L. (1957), *Introduction to Operations Research*, John Wiley & Sons, New York.

Currie, R. M. (1960), *Work Study*, Pitman, London, U.K.

Dandy, G. C., Walker, D., Daniell, T., and Warner, R. F. (2008), *Planning and Design of Engineering Systems*, 2nd edn., Taylor & Francis, London, U.K.

Dell'Isola, A. J. (1982), *Value Engineering in the Construction Industry*, Construction Publishing Corp., New York.

Gelb, A. (ed.) (1974), *Applied Optimal Estimation*, MIT Press, Cambridge, MA.

Gleick, J. (1988), *Chaos*, Sphere Books, London, U.K., pp. 278–279.

Graupe, D. (1976), *Identification of Systems*, Krieger, New York.

Gross, D. and Harris, C. M. (1974), *Fundamentals of Queuing Theory*, John Wiley & Sons, New York.

Hall, A. D. (1962), *A Methodology for Problem Solving*, D. Van Nostrand, Princeton, NJ.

Hall, A. D. and Dracup, J. A. (1970), *Water Resources Systems Engineering*, McGraw-Hill, New York.

Heller, J. (1964), *Catch-22*, Corgi Books, London, U.K.

Hicks, H. G. and Gullett, C. R. (1976), *The Management of Organisations*, McGraw-Hill, New York.

Hillier, F. S. and Lieberman, G. J. (1990), *Introduction to Operations Research*, 5th edn., McGraw-Hill, New York.

Hilmer, F. G. and Donaldson, L. (1996), *Management Redeemed*, The Free Press, New York.

Hollick, M. (1993), *An Introduction to Project Evaluation*, Longman Cheshire, Sydney, New South Wales, Australia.

Hsia, T. C. (1977), *System Identification*, Lexington Books, Lexington, MA.

Huff, D. (1954), *How to Lie with Statistics*, Victor Gollancz Limited, London, U.K.

International Labour Office (ILO) (1969), *Introduction to Work Study*, ILO, Geneva, Switzerland.

Jewell, T. K. (1986), *A Systems Approach to Civil Engineering Planning and Design*, Harper & Row, New York.

Klir, G. J. (1969), *An Approach to General Systems Theory*, Van Nostrand Reinold, New York.

Kramer, N. J. T. A. and de Smit, J. (1977), *Systems Thinking*, Martinus Nijhoff, Leiden, the Netherlands.

Lehmann, D. R. (1979), *Market Research and Analysis*, Irwin, Homewood, IL.

Meadows, D. H., Meadows, D. L., Randers, J., and Behrens, W. W. (1972), The limits to growth, A report for the Club of Rome's project on the predicament of mankind, Pan, London, U.K.

Meredith, D. D., Wong, K. W., Woodhead, R. W., and Wortman, R. H. (1985), *Design and Planning of Engineering Systems*, 2nd edn., Prentice Hall, Englewood Cliffs, NJ.

de Neufville, R. (1990), *Applied Systems Analysis: Engineering Planning and Technology Management*, McGraw-Hill, New York.

de Neufville, R. and Stafford, J. H. (1971), *Systems Analysis for Engineers and Managers*, McGraw-Hill, New York.

Ossenbruggen, P. J. (1984), *Systems Analysis for Civil Engineers*, John Wiley & Sons, New York.

Parkinson, N. C. (1957), *Parkinson's Law or the Pursuit of Progress*, John Murray, London, U.K.

Parkinson, N. C. (1968), *Applying to Married Women of the Western World—Mrs Parkinson's Law*, Cox & Wyman Ltd, London, U.K.

Porter, A. (1969), *Cybernetics Simplified*, The English Universities Press, London, U.K.

Sage, A. P. and Melsa, J. (1971), *System Identification*, Academic Press, New York.

Sage, A. P. and White, C. C. (1977), *Optimum Systems Control*, Prentice Hall, Englewood Cliffs, NJ.

Stanton, W. J., Miller, K. E., and Layton, R. A. (1985), *Fundamentals of Marketing*, 1st Australian edition, McGraw-Hill, Sydney, New South Wales, Australia.

Stark, R. M. and Nicholls, R. L. (1972), *Mathematical Foundations for Design – Civil Engineering Systems*, McGraw-Hill, New York.

Stoner, J. A. F., Collins, R. R., and Yetton, P. W. (1985), *Management in Australia*, Prentice Hall, Sydney, New South Wales, Australia.

Vincent, P. (1977), *The Two Ronnies—Nice to Be With You Again!*, Star Book, London, U.K.

Wilmut, R. (1982), *No More Curried Eggs for Me*, Methuen, London, U.K., pp. 47–49.

Index

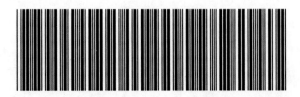